南京航空航天大学研究生系列精品教材

图像处理技术及其应用

王开福 著

科学出版社

北京

内 容 简 介

本书是南京航空航天大学研究生教育优秀工程(二期)建设项目。全书主要内容包括图像运算、图像变换、图像降噪、图像增强、图像分割、图像恢复与再现、图像配准与相关以及图像形态运算等。

本书可作为高等院校航空宇航科学与技术、力学、机械工程、光学工程、材料科学与工程、电子科学与技术、信息与通信工程、计算机科学与技术等专业的研究生教材，也可供理工科相关专业的高校教师、研究人员和技术人员参考。

图书在版编目(CIP)数据

图像处理技术及其应用/王开福著. —北京：科学出版社，2015.6
ISBN 978-7-03-044568-1

Ⅰ. ①图… Ⅱ. ①王… Ⅲ. ①图形软件－高等学校－教材
Ⅳ. ①TP391.41

中国版本图书馆 CIP 数据核字(2015)第 124260 号

责任编辑：余 江 张丽花 / 责任校对：包志虹
责任印制：徐晓晨 / 封面设计：迷底书装

科学出版社 出版
北京东黄城根北街 16 号
邮政编码：100717
http://www.sciencep.com

北京建宏印刷有限公司 印刷
科学出版社发行 各地新华书店经销
*

2015 年 6 月第 一 版 开本：787×1092 1/16
2019 年 1 月第三次印刷 印张：11 1/4
字数：267 000

定价：45.00 元
(如有印装质量问题，我社负责调换)

前　言

　　图像处理分为模拟图像处理和数字图像处理。模拟图像处理是指采用光学方法对模拟图像进行图像处理，而数字图像处理则是指采用数字方法对数字图像进行图像处理。模拟图像只能采用光学方法进行图像处理，不能直接采用数字方法进行图像处理，然而模拟图像通过采样和量化变成数字图像后，即可通过数字方法进行图像处理，因此通常所说的图像处理是指数字图像处理。

　　本书是南京航空航天大学研究生教育优秀工程(二期)建设项目，在阐述数字图像处理的基本原理和基础理论的基础上，重点论述数字图像处理的最新技术及其实际应用。全书主要内容包括图像运算、图像变换、图像降噪、图像增强、图像分割、图像恢复与再现、图像配准与相关、图像形态运算等。图像运算部分在介绍图像几何和算术运算的基础上，着重讨论图像卷积与相关的理论与应用；图像变换部分既讨论经典的傅里叶变换和余弦变换，又阐述最新的小波变换技术及其应用；图像降噪部分对空域平滑和频域低通，尤其是同态低通和小波低通，等滤波降噪技术及其应用进行详细论述；直方图变换、灰度变换、空域滤波和频域滤波等常用图像增强技术及其应用在图像增强部分分别进行详细讨论；图像分割部分主要论述阈值分割、边缘检测和边界跟踪的原理和方法；图像恢复与再现部分，在分析图像恢复原理和算法的基础上，详细论述数字全息的再现技术及其应用；图像配准的原理和算法，特别是数字散斑相关技术及其应用，在图像配准与相关部分分别进行详细论述；图像形态运算部分对腐蚀与膨胀、开启与闭合等形态运算分别进行论述。

　　由于作者水平有限和编写时间仓促，若存在不当之处，敬请批评指正。

作　者
2015 年 3 月于南京航空航天大学

目　　录

前言

第1章　图像处理基础 ··· 1
1.1　图像概念 ·· 1
1.1.1　图像及其分类 ·· 1
1.1.2　采样与量化 ··· 1
1.2　图像软件 ·· 2
1.3　图像表示 ·· 3
1.3.1　像素坐标 ··· 4
1.3.2　空间坐标 ··· 4
1.4　图像输入与输出 ·· 5
1.4.1　图像读入 ··· 5
1.4.2　图像写出 ··· 5
1.4.3　图像显示 ··· 6
1.5　数据类型及其转换 ··· 6
1.5.1　数据类型 ··· 6
1.5.2　数据类型转换 ·· 6
1.6　图像类型及其转换 ··· 7
1.6.1　图像类型 ··· 7
1.6.2　图像类型转换 ·· 8
1.7　颜色模型及其转换 ··· 10
1.7.1　RGB 模型 ·· 10
1.7.2　HSV 模型及其转换 ·· 11
1.7.3　YIQ/NTSC 模型及其转换 ·· 12
1.7.4　YCbCr 模型及其转换 ·· 13
1.8　图像格式及其转换 ··· 14

第2章　图像运算 ·· 15
2.1　几何运算 ·· 15
2.1.1　图像插值 ··· 15
2.1.2　图像平移 ··· 17
2.1.3　图像缩放 ··· 18
2.1.4　图像旋转 ··· 19
2.1.5　图像剪切 ··· 19

2.2 算术运算 ··· 20
2.2.1 图像相加 ··· 20
2.2.2 图像相减 ··· 22
2.2.3 绝对差值 ··· 23
2.2.4 图像相乘 ··· 24
2.2.5 图像相除 ··· 25
2.2.6 图像线性组合 ··· 25
2.3 卷积与相关运算 ··· 26
2.3.1 图像卷积 ··· 26
2.3.2 图像相关 ··· 26
2.3.3 卷积和相关应用 ··· 26

第3章 图像变换 ··· 29
3.1 傅里叶变换 ··· 29
3.1.1 连续傅里叶变换 ··· 29
3.1.2 傅里叶变换性质 ··· 30
3.1.3 离散傅里叶变换 ··· 32
3.1.4 快速傅里叶变换 ··· 33
3.1.5 离散傅里叶变换算法 ··· 35
3.1.6 离散傅里叶变换应用 ··· 39
3.2 余弦变换 ··· 40
3.2.1 离散余弦变换 ··· 40
3.2.2 离散余弦变换算法 ··· 41
3.2.3 离散余弦变换应用 ··· 43
3.3 小波变换 ··· 43
3.3.1 连续小波变换 ··· 44
3.3.2 离散小波变换 ··· 45
3.3.3 离散小波变换算法 ··· 47
3.3.4 离散小波变换应用 ··· 50

第4章 图像降噪 ··· 51
4.1 空域平滑滤波 ··· 51
4.1.1 均值滤波 ··· 51
4.1.2 中值滤波 ··· 53
4.1.3 自适应滤波 ··· 55
4.2 频域低通滤波 ··· 56
4.2.1 理想低通滤波 ··· 56
4.2.2 巴特沃斯低通滤波 ··· 59
4.2.3 指数低通滤波 ··· 60

4.3 同态低通滤波···62
　　4.3.1 同态低通滤波原理···62
　　4.3.2 同态低通滤波应用···63
4.4 小波低通滤波···64
　　4.4.1 小波低通滤波原理···64
　　4.4.2 小波低通滤波算法···64
　　4.4.3 小波低通滤波应用···65

第5章 图像增强···66
5.1 直方图变换···66
　　5.1.1 直方图···66
　　5.1.2 直方图均衡化···68
　　5.1.3 直方图自适应均衡化···69
5.2 灰度变换··70
　　5.2.1 线性变换···70
　　5.2.2 非线性变换···71
　　5.2.3 灰度变换算法···72
5.3 空域滤波··73
　　5.3.1 空域平滑滤波···73
　　5.3.2 空域锐化滤波···74
5.4 频域滤波··76
　　5.4.1 频域低通滤波···76
　　5.4.2 频域高通滤波···76

第6章 图像分割···80
6.1 阈值分割··80
　　6.1.1 阈值分割原理···80
　　6.1.2 阈值确定方法···81
　　6.1.3 阈值分割应用···83
6.2 边缘检测与边界跟踪··83
　　6.2.1 边缘检测···84
　　6.2.2 边界跟踪···87

第7章 图像恢复与再现··89
7.1 图像恢复··89
　　7.1.1 图像恢复原理···89
　　7.1.2 图像恢复算法···91
　　7.1.3 图像恢复应用···95
7.2 图像再现··96
　　7.2.1 数字全息再现原理···96

第8章　图像配准与相关 ·············· 99
8.1　图像配准 ·············· 99
- 8.1.1　图像配准原理 ·············· 99
- 8.1.2　图像配准算法 ·············· 99
- 8.1.3　图像配准应用 ·············· 100

8.2　图像相关 ·············· 101
- 8.2.1　数字散斑相关原理 ·············· 101
- 8.2.2　数字散斑相关应用 ·············· 103

第9章　图像形态运算 ·············· 105
9.1　集合概念 ·············· 105
9.2　结构元素 ·············· 105
- 9.2.1　结构元素形成 ·············· 105
- 9.2.2　结构元素分解 ·············· 107

9.3　膨胀和腐蚀 ·············· 108
- 9.3.1　膨胀和腐蚀原理 ·············· 108
- 9.3.2　图像边界处理 ·············· 109
- 9.3.3　膨胀和腐蚀算法 ·············· 109
- 9.3.4　膨胀和腐蚀应用 ·············· 110

9.4　开启和闭合 ·············· 111
- 9.4.1　开启和闭合原理 ·············· 111
- 9.4.2　开启和闭合算法 ·············· 112
- 9.4.3　开启和闭合应用 ·············· 113

附录 I　常用基本函数 ·············· 114
I.1　Language Fundamentals ·············· 114
- I.1.1　Entering Commands ·············· 114
- I.1.2　Matrices and Arrays ·············· 114
- I.1.3　Operators and Elementary Operations ·············· 116
- I.1.4　Special Characters ·············· 117
- I.1.5　Data Types ·············· 117
- I.1.6　Dates and Time ·············· 125

I.2　Mathematics ·············· 125
- I.2.1　Elementary Math ·············· 125
- I.2.2　Linear Algebra ·············· 130
- I.2.3　Statistics and Random Numbers ·············· 132
- I.2.4　Interpolation ·············· 133
- I.2.5　Fourier Analysis and Filtering ·············· 134

7.2.2　数字全息再现应用 ·············· 97

I.3 Graphics ··· 134
I.3.1 2-D and 3-D Plots ··· 134
I.3.2 Formatting and Annotation ··· 138
I.3.3 Images ··· 140
I.3.4 Printing and Exporting ··· 140
I.3.5 Graphics Objects ··· 141
I.4 Programming Scripts and Functions ··· 143
I.4.1 Control Flow ··· 143
I.4.2 Scripts ··· 143
I.4.3 Functions ··· 143
I.4.4 Debugging ··· 144
I.4.5 Coding and Productivity Tips ··· 145
I.4.6 Programming Utilities ··· 145
I.5 Data and File Management ··· 145
I.5.1 Workspace Variables ··· 145
I.5.2 Data Import and Export ··· 146

附录 II 图像处理函数 ··· 148
II.1 Import, Export, and Conversion ··· 148
II.1.1 Basic Import and Export ··· 148
II.1.2 Scientific File Formats ··· 148
II.1.3 High Dynamic Range Images ··· 148
II.1.4 Large Image Files ··· 148
II.1.5 Image Type Conversion ··· 149
II.1.6 Synthetic Images ··· 149
II.2 Display and Exploration ··· 149
II.2.1 Basic Display ··· 149
II.2.2 Interactive Exploration with the Image Viewer App ··· 150
II.2.3 Build Interactive Tools ··· 150
II.3 Geometric Transformation, Spatial Referencing, and Image Registration ··· 151
II.3.1 Geometric Transformations ··· 151
II.3.2 Spatial Referencing ··· 152
II.3.3 Automatic Registration ··· 152
II.3.4 Control Point Registration ··· 153
II.4 Image Enhancement ··· 153
II.4.1 Contrast Adjustment ··· 153
II.4.2 Image Filtering ··· 154
II.4.3 Morphological Operations ··· 154

	II.4.4	Deblurring	155
	II.4.5	ROI-Based Processing	155
	II.4.6	Neighborhood and Block Processing	156
	II.4.7	Image Arithmetic	156
II.5	Image Analysis		156
	II.5.1	Object Analysis	156
	II.5.2	Region and Image Properties	157
	II.5.3	Texture Analysis	157
	II.5.4	Image Quality	158
	II.5.5	Image Segmentation	158
	II.5.6	Image Transforms	158
II.6	Color		158
II.7	Code Generation		159

附录 III 小波分析函数 161

III.1	Wavelets and Filter Banks		161
	III.1.1	Real and Complex-Valued Wavelets	161
	III.1.2	Orthogonal and Biorthogonal Filter Banks	161
	III.1.3	Lifting	162
	III.1.4	Wavelet Design	162
III.2	Continuous Wavelet Analysis		163
III.3	Discrete Wavelet Analysis		163
	III.3.1	Signal Analysis	163
	III.3.2	Image Analysis	164
	III.3.3	3-D Analysis	165
	III.3.4	Multisignal Analysis	166
III.4	Wavelet Packet Analysis		166
III.5	Denoising		168

参考文献 169

第 1 章　图像处理基础

1.1　图像概念

1.1.1　图像及其分类

通常所说的图像(image)，其含义十分广泛。图像既指艺术领域人或物的复制，如画像和塑像；也指光学领域人或物的复制，如镜像和影像；还指数学领域二维或多维数组的映射，如图形和图片；甚至指并不存在的人或物的反映，等等。图像处理中所说的图像主要是指光学领域人或物的复制以及数学领域二维或多维数组的映射。

根据人眼视觉特性，图像分为可见图像(visible image)和不可见图像(invisible image)。人眼能够感知的图像称为可见图像，如照片(单幅图像)和电影(序列图像)等；反之，人眼不能感知的图像则称为不可见图像，如电磁波谱和温度分布等。不可见图像通常可以转化为可见图像，如红外热像技术可以把温度分布转变为可见图像。

根据坐标和灰度是否连续，图像分为模拟图像(analogue image)和数字图像(digital image)。模拟图像是指坐标和灰度都具有连续性，如采用照相底片记录的照片；数字图像是指坐标和灰度均具有离散性，如采用数码相机拍摄的照片。

1.1.2　采样与量化

模拟图像只能采用光学方法进行处理，而不能直接采用数字方法进行处理，但是模拟图像通过采样和量化变成数字图像后，即可通过数字方法进行处理。

1. 采样

所谓采样(sampling)是指将模拟图像的连续空间坐标离散化为离散空间坐标。设对模拟图像 $A(x,y)$ 进行均匀采样，在 x,y 方向的采样间隔分别为 $\Delta x, \Delta y$，则采样后的数字图像可表示为

$$B(m,n) = A(x,y)S(m,n) = \sum_{m=0}^{M-1}\sum_{n=0}^{N-1} A(x,y)\delta(x-m\Delta x, y-n\Delta y) \tag{1.1}$$

式中

$$S(m,n) = \sum_{m=0}^{M-1}\sum_{n=0}^{N-1} \delta(x-m\Delta x, y-n\Delta y) \tag{1.2}$$

为采样函数，如图 1.1 所示。

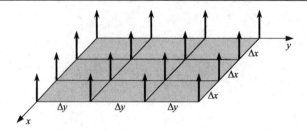

图 1.1　采样函数

如果采样点取得彼此足够靠近，那么采样数据就是原来图像的精确表示，即通过插值就能精确再现原来图像。根据采样定理(sampling theorem)，即香农(Shannon)或奈奎斯特(Nyquist)采样定理，采样频率必须高于模拟图像最高频率的 2 倍，才不至于在采样过程中产生混叠效应(频谱交叉)。因此采样间隔需要满足

$$\Delta x \leqslant \frac{1}{2u_{\max}}, \quad \Delta y \leqslant \frac{1}{2v_{\max}} \tag{1.3}$$

式中，$u_{\max},v_{\max}$ 分别为模拟图像在 x,y 方向的最大空间频率。

2. 量化

所谓量化(quantizing)是指把模拟图像的连续灰度分布离散化为离散灰度分布，如二值图像(binary image)的灰度级为 $2^1=2$，其每个像素的灰度值为 0(黑)或 1(白)；8 位无符号整型灰度图像(grayscale image)的灰度级为 $2^8=256$，其每个像素的灰度值为 0(黑)、1,2,…,254 或 255(白)。

1.2　图　像　软　件

MATLAB(matrix laboratory)是 MathWorks 公司开发的面向科学和工程计算的高级编程语言。MATLAB 语言具有编程简单和易学易用等优点。目前，MATLAB 的图像处理工具箱(Image Processing Toolbox)和小波分析工具箱(Wavelet Toolbox)已广泛应用于图像处理。

目前，MATLAB 的常用版本如图 1.2 所示。

图 1.2　MATLAB R2014a

图 1.3 所示为 MATLAB 命令窗口。命令窗口用于输入 MATLAB 语句。

图 1.3　MATLAB 命令窗口

图 1.4 所示为 MATLAB 编辑窗口。编辑窗口用于编写扩展名为 m 的 M 文件。M 文件是文本文件，通常包含两种类型：脚本(script)和函数(function)。

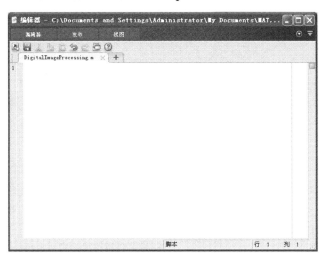

图 1.4　MATLAB 编辑窗口

1.3　图 像 表 示

MATLAB 的基本数据结构是数组，数组就是一组实数或复数的有序集合。而图像正是灰度(或颜色)数据的实值有序集合，因此 MATLAB 非常适合表征图像。

MATLAB 把灰度图像存储为二维数组(矩阵)，数组元素对应图像像素。如 M 行 N 列的灰度图像可表示为

$$I(:,:) = \begin{bmatrix} I(1,1) & I(1,2) & \cdots & I(1,n) & \cdots & I(1,N) \\ I(2,1) & I(2,2) & \cdots & I(2,n) & \cdots & I(2,N) \\ \vdots & \vdots & & \vdots & & \vdots \\ I(m,1) & I(m,2) & \cdots & I(m,n) & \cdots & I(m,N) \\ \vdots & \vdots & & \vdots & & \vdots \\ I(M,1) & I(M,2) & \cdots & I(M,n) & \cdots & I(M,N) \end{bmatrix} \quad (1.4)$$

式中，(m,n) 和 $I(m,n)$ 分别为图像坐标和灰度。

真彩图像在 MATLAB 中存储为三维数组，其中沿第三维方向的第 1 个面表示红色分量，第 2 个面表示绿色分量，第 3 个面表示蓝色分量。如 M 行 N 列的真彩图像的 3 个分量可分别表示为

$$I(:,:,i) = \begin{bmatrix} I(1,1,i) & I(1,2,i) & \cdots & I(1,n,i) & \cdots & I(1,N,i) \\ I(2,1,i) & I(2,2,i) & \cdots & I(2,n,i) & \cdots & I(2,N,i) \\ \vdots & \vdots & & \vdots & & \vdots \\ I(m,1,i) & I(m,2,i) & \cdots & I(m,n,i) & \cdots & I(m,N,i) \\ \vdots & \vdots & & \vdots & & \vdots \\ I(M,1,i) & I(M,2,i) & \cdots & I(M,n,i) & \cdots & I(M,N,i) \end{bmatrix},\ (i=1,2,3) \quad (1.5)$$

式中，(m,n,i) 和 $I(m,n,i)$ 分别为第 i 面的图像坐标和灰度。

1.3.1 像素坐标

在 MATLAB 中，像素被看成离散点，像素坐标只能取离散正整数值，坐标排序从上到下，从左到右，如图 1.5 所示。

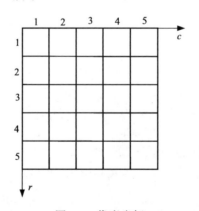

图 1.5 像素坐标

像素坐标与 MATLAB 数组坐标具有一一对应关系，因此通过数组坐标可以读写图像像素值。如 MATLAB 中坐标为 (3, 2) 的数组元素对应于第 3 行第 2 列的图像像素。

1.3.2 空间坐标

在空间坐标中，像素位置可用连续坐标 (x, y) 表示，如图 1.6 所示。在 MATLAB 中，x 坐标向右为正，y 坐标向下为正。

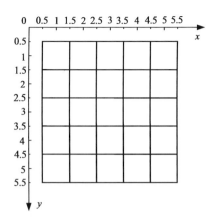

图 1.6 空间坐标

空间坐标与像素坐标之间具有对应关系,如像素坐标与像素中心所对应的空间坐标完全相同。然而,两种坐标之间也存在差别,如在像素坐标中左上角的坐标是(1,1),而在空间坐标中左上角的坐标是(0.5, 0.5),引起这种差别的主要原因是像素坐标具有离散性,而空间坐标具有连续性。另外,在像素坐标中左上角的坐标始终是(1, 1),而在空间坐标中左上角的坐标可以指定任何值。

1.4 图像输入与输出

1.4.1 图像读入

MATLAB 利用 imread 函数把图像从图像文件读到工作空间,其主要用法如下:

(1) A = imread(filename, format),把灰度图像或彩色图像从图像文件读到工作空间。A 为包含图像数据的数组。对于灰度图像,则 A 是 $M \times N$ 数组;对于真彩图像,则 A 是 $M \times N \times 3$ 数组;对于采用 CMYK 颜色空间表示的 TIFF 格式彩色图像,则 A 是 $M \times N \times 4$ 数组。

(2) [X, map] = imread(…),把索引图像从图像文件读到工作空间,其中 X 是索引数组,map 是颜色矩阵。

1.4.2 图像写出

MATLAB 利用 imwrite 函数把图像从工作空间写到图像文件,其主要用法如下:

(1) imwrite(A, filename, format),把数组 A 从工作空间写到图像文件。图像文件名为 filename、格式由 format 指定。A 可以是 $M \times N$(灰度图像)或 $M \times N \times 3$(真彩图像)数组。对于 TIFF 文件,A 可以是采用 CMYK 颜色空间的 $M \times N \times 4$ 数组。

(2) imwrite(X, map, filename, format),把索引图像 X 及其颜色矩阵 map 从工作空间写到文件名为 filename 和格式为 format 的图像文件。

1.4.3 图像显示

MATLAB 利用 imshow 函数显示图像，其主要用法如下：

(1) imshow(I)，在图像窗口显示灰度图像 I。

(2) imshow(I, [low high])，在图像窗口显示灰度图像 I，并指定灰度显示范围为[low high]。灰度等于或小于 low 值时显示为黑，灰度等于或大于 high 值时显示为白。如果用空矩阵[ ]代替[low high]，则 imshow 将在最小灰度值和最大灰度值之间显示图像。

(3) imshow(RGB)，显示真彩图像 RGB。

(4) imshow(BW)，显示二值图像 BW。像素值 0 和 1 分别显示为黑和白。

(5) imshow(X, map)，借助颜色矩阵 map 显示索引图像 X。颜色矩阵可以有任意行（但只有 3 列），每一行代表一种颜色，每行的 3 个元素分别表示红、绿和蓝，颜色值位于[0.0,1.0]范围。

(6) imshow(filename)，显示图像文件，其中 imshow 函数将通过调用 imread 函数读取图像文件，但并不把图像数据读到 MATLAB 工作空间。

1.5 数据类型及其转换

1.5.1 数据类型

在 MATLAB 中，灰度图像和真彩图像的数据可以是 8 位无符号整型、16 位无符号整型、16 位带符号整型、单精度浮点型、双精度浮点型或逻辑型；索引图像可以是 8 位无符号整型、16 位无符号整型、双精度浮点型或逻辑型；二值图像只能是逻辑型。

1.5.2 数据类型转换

利用 MATLAB 提供的函数可以进行数据类型转换。

1. MATLAB 利用 im2uint8 函数把图像数据转换为 8 位无符号整型，其主要用法如下

(1) A = im2uint8(I)，灰度图像转换为 8 位无符号整型；

(2) A = im2uint8(RGB)，真彩图像转换为 8 位无符号整型；

(3) A = im2uint8(BW)，二值图像转换为 8 位无符号整型；

(4) A = im2uint8(X, 'indexed')，索引图像转换为 8 位无符号整型。

2. MATLAB 利用 im2uint16 函数把图像数据转换为 16 位无符号整型，其主要用法如下

(1) A = im2uint16(I)，灰度图像转换为 16 位无符号整型；

(2) A = im2uint16(RGB)，真彩图像转换为 16 位无符号整型；

(3) A = im2uint16(BW)，二值图像转换为 16 位无符号整型；

(4) A = im2uint16(X, 'indexed')，索引图像转换为 16 位无符号整型。

3. MATLAB 利用 im2int16 函数把图像数据转换为 16 位带符号整型，其主要用法如下

(1) A = im2int16(I)，灰度图像转换为 16 位带符号整型；

(2) A = im2int16(RGB)，真彩图像转换为 16 位带符号整型；

(3) A = im2int16(BW)，二值图像转换为 16 位带符号整型。

4. MATLAB 利用 im2single 函数把图像数据转换为单精度浮点型，其主要用法如下

(1) A = im2single(I)，灰度图像转换为单精度浮点型；

(2) A = im2single(RGB)，真彩图像转换为单精度浮点型；

(3) A = im2single(BW)，二值图像转换为单精度浮点型；

(4) A = im2single(X, 'indexed')，索引图像转换为单精度浮点型。

5. MATLAB 利用 im2double 函数把图像数据转换为双精度浮点型，其主要用法如下

(1) A = im2double(I)，灰度图像转换为双精度浮点型；

(2) A = im2double(RGB)，真彩图像转换为双精度浮点型；

(3) A = im2double(BW)，二值图像转换为双精度浮点型；

(4) A = im2double(X, 'indexed')，索引图像转换为双精度浮点型。

1.6 图像类型及其转换

1.6.1 图像类型

1. 二值图像

二值图像以逻辑数组存储，每个像素取值为 0（黑）或 1（白），如图 1.7 所示。

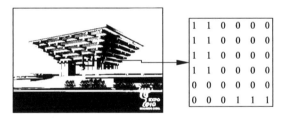

图 1.7 二值图像及其像素值

2. 灰度图像

灰度图像由一个数据矩阵组成，数据矩阵中的元素值表示像素灰度或亮度，如图 1.8 所示。矩阵数据可以是 8 位无符号整型、16 位无符号整型、16 位带符号整型、单精度浮点型或双精度浮点型。

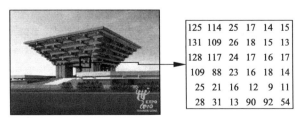

图 1.8 灰度图像及其像素值

对单精度浮点型或双精度浮点型，灰度 0 表示黑，1 表示白。对 8 位无符号整型、16 位无符号整型或 16 位带符号整型，最小灰度表示黑，最大灰度表示白。

3. 索引图像

索引图像由一个索引数组和一个颜色矩阵组成。索引数组的像素值是颜色矩阵的索引。颜色矩阵是 3 列数组，其元素值为[0, 1]之间的双精度浮点型，每一行的 3 个元素分别表示红、绿和蓝。像素颜色由像素值所对应的三个颜色值确定。

如果索引数组是单精度型或双精度型，像素值 1 指向颜色矩阵中的第一行，2 指向第二行，以此类推。如果索引数组是逻辑型、8 位无符号整型或 16 位无符号整型，像素值 0 指向颜色值中的第一行，1 指向第二行，以此类推，如图 1.9 所示。

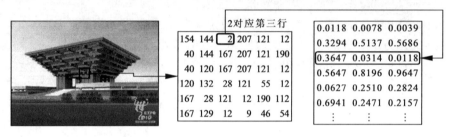

图 1.9　索引图像及其像素值到颜色值的映射

4. 真彩图像

真彩图像的像素由红、绿和蓝 3 种颜色值构成。MATLAB 把真彩图像存储为三维数组，第三维方向是由红、绿和蓝 3 个颜色面组成，像素颜色由 3 种颜色值的组合确定。

MATLAB 把真彩图像存储为 24 位图像，红、绿和蓝各占 8 位，如图 1.10 所示。真彩图像可以是 8 位无符号整型、16 位无符号整型、单精度浮点型或双精度浮点型。单精度浮点型或双精度浮点型真彩图像，颜色分量(0, 0, 0)显示黑，(1, 1, 1)显示白。

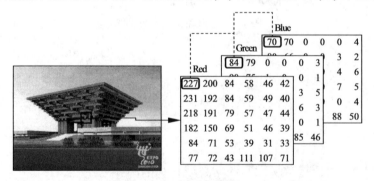

图 1.10　真彩图像及其颜色面

1.6.2　图像类型转换

在 MATLAB 中，一种类型的图像可以转换为另一种类型的图像。例如，如果要对索引图像进行滤波,则首先必须要把索引图像转换为真彩图像。MATLAB 提供了各种类型图像的转换函数。

1. 灰度、索引和真彩图像转换为二值图像

MATLAB 利用 im2bw 函数把灰度、索引和真彩图像转换为二值图像,其主要用法如下:

(1) A = im2bw(I, level),灰度图像转换为二值图像。输入图像中灰度值大于 level(level 在[0, 1]范围取值)的所有像素在输出图像中的灰度值都取 1(白),其余灰度值都取 0(黑)。

(2) A = im2bw(X, map, level),索引图像转换为二值图像。

(3) A = im2bw(RGB, level),真彩图像转换为二值图像。

当输入图像不是灰度图像时,im2bw 函数先把图像转换为灰度图像,然后再把灰度图像转换为二值图像。

2. 索引图像转换为灰度图像

MATLAB 利用 ind2gray 函数把索引图像转换为灰度图像,其主要用法如下:

A = ind2gray(X, map),索引图像转换为灰度图像,其中 ind2gray 函数将丢失色调(hue)和饱和度(saturation)信息,而仅保留亮度(luminance)信息。

3. 真彩图像转换为灰度图像

MATLAB 利用 rgb2gray 函数把真彩图像转换为灰度图像,其主要用法如下:

A = rgb2gray(RGB),真彩图像转换为灰度图像,其中 rgb2gray 函数将丢掉色调和饱和度信息,而仅保留亮度信息。

4. 灰度和二值图像转换为索引图像

MATLAB 利用 gray2ind 函数把灰度和二值图像转换为索引图像,其主要用法如下:

(1) [X, map] = gray2ind(I, n),灰度图像转换为索引图像,其中 n 为颜色矩阵行数,在 1×2^{16} 之间取值,缺省时 n = 64。

(2) [X, map] = gray2ind(BW, n),二值图像转换为索引图像,其中 n 为颜色矩阵行数,缺省时 n = 2。

5. 索引图像转换为真彩图像

MATLAB 利用 ind2rgb 函数把索引图像转换为真彩图像,其主要用法如下:

A = ind2rgb(X, map),索引图像转换为真彩图像。

6. 真彩图像转换为索引图像

MATLAB 利用 rgb2ind 函数把真彩图像转换为索引图像,其主要用法如下:

[X, map] = rgb2ind(RGB, n),真彩图像转换为索引图像,其中 n 指定颜色数量,其值小于或等于 65536。

7. 矩阵转换为灰度图像

MATLAB 利用 mat2gray 函数把矩阵转换为灰度图像,其主要用法如下:

A = mat2gray(M, [amin amax])，数据矩阵转换为灰度图像，其中，A 的取值范围在 0.0(黑)和 1.0(白)之间，[amin amax]是参照 M 而指定的数值范围，其分别对应 A 中的 0.0 和 1.0，当缺省时，[amin amax]分别为 M 中的最小值和最大值。

1.7 颜色模型及其转换

颜色由色调(hue)、饱和度(saturation)和亮度(luminance)等因素确定。色调与光的主波长有关。饱和度与单色光中掺入的白光程度有关，单色光的饱和度为 100%，白光的饱和度为 0。亮度与光的强度有关。色调和饱和度的集合称为色度。颜色可用亮度和色度进行表示。

常用的 MATLAB 颜色模型包括 RGB、HSV、YIQ/NTSC 和 YCbCr 等模型。

1.7.1 RGB 模型

RGB 模型通过红(red)、绿(green)和蓝(blue)等三种基本色调(三基色)的相加混合进行颜色描述，如图 1.11 所示。

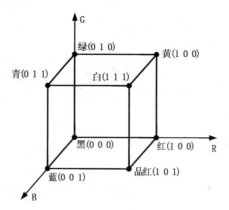

图 1.11 RGB 模型

不同比例的三基色相加混合可以产生其他颜色，因此 RGB 模型称为加色系统。

图 1.12 所示为 RGB 图像及其三个分量图像。图 1.12(a)为 RGB 图像，图 1.12(b)、图 1.12(c)和图 1.12(d)分别为红、绿和蓝分量图像。

(a)　　　　　　　　　(b)

图 1.12 RGB 图像及其三个分量图像

(c)　　　　　　　　　　　(d)

图 1.12　RGB 图像及其三个分量图像(续)

1.7.2　HSV 模型及其转换

HSV 模型通过色调(hue)、饱和度(saturation)和亮度(value)等三个分量进行颜色描述。

图 1.13 所示为 HSV 图像及其三个分量图像。图 1.13(a)为 HSV 图像，图 1.13(b)、图 1.13(c)和图 1.13(d)分别为色调、饱和度和亮度分量图像。

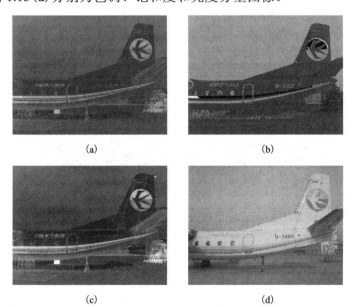

图 1.13　HSV 图像及其三个分量图像

RGB 图像的 R、G、B 分量与 HSV 图像的 H、S、V 分量之间的转换关系可表示为

$$V = \max(R,G,B)$$
$$S_1 = V - \min(R,G,B), S = S_1/V \tag{1.6}$$
$$H_1 = \begin{cases} (G-B)/(6S_1) & (R=V) \\ [2+(B-R)/S_1]/6 & (G=V) \\ [4+(R-G)/S_1]/6 & (B=V) \end{cases} \quad H = \begin{cases} H_1+1 & (H_1<0) \\ H_1 & (\text{其他}) \end{cases}$$

式中，$\max(\cdots)$ 和 $\min(\cdots)$ 分别表示取最大和最小值，R, G, B 取值范围均为 [0, 1]。

MATLAB 采用 rgb2hsv 函数把 RGB 图像转换为 HSV 图像或采用 hsv2rgb 函数把 HSV 图像转换为 RGB 图像，其主要用法如下：

(1) HSV = rgb2hsv(RGB)，把 RGB 图像转换为 HSV 图像，其中 RGB 和 HSV 都是 $M \times N \times 3$ 矩阵。RGB 和 HSV 的元素值都处于[0,1]范围。

RGB 模型在第三维方向的三个面分别表示红、绿和蓝分量，而 HSV 模型在第三维方向的三个面分别表示色调、饱和度和亮度分量。

(2) RGB = hsv2rgb(HSV)，把 HSV 图像转换为 RGB 图像。

1.7.3 YIQ/NTSC 模型及其转换

YIQ/NTSC 模型通过亮度、色调和饱和度等三个分量进行颜色描述。亮度 Y 表示灰度信息，而色调 I 和饱和度 Q 表示色度(chrominance)信息。

图 1.14 所示为 NTSC 图像及其三个分量图像。图 1.14(a) 为 NTSC 图像，图 1.14(b)、图 1.14(c) 和图 1.14(d) 分别为亮度(灰度)、色调和饱和度分量图像。

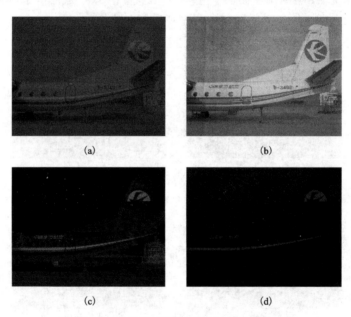

图 1.14　NTSC 图像及其三个分量图像

RGB 图像的 R、G、B 分量与 NTSC 图像的 Y、I、Q 分量之间的转换关系可表示为

$$\begin{bmatrix} Y \\ I \\ Q \end{bmatrix} = \begin{bmatrix} 0.2989 & 0.5870 & 0.1140 \\ 0.5959 & -0.2744 & -0.3216 \\ 0.2115 & -0.5229 & 0.3114 \end{bmatrix} \begin{bmatrix} R \\ G \\ B \end{bmatrix} \quad (1.7)$$

MATLAB 采用 rgb2ntsc 函数把 RGB 图像转换为 NTSC 图像或采用 ntsc2rgb 函数把 NTSC 图像转换为 RGB 图像，其主要用法如下：

(1) YIQ = rgb2ntsc(RGB)，把 RGB 图像转换为 NTSC 图像，其中 RGB 和 YIQ 都是 $M \times N \times 3$ 矩阵。RGB 和 YIQ 的元素值都处于[0,1]范围。

(2) RGB = ntsc2rgb(YIQ)，把 NTSC 图像转换为 RGB 图像。

在 MATLAB 中，rgb2gray 函数和 ind2gray 函数都是通过利用 rgb2ntsc 函数把彩色图像转换为灰度图像。

1.7.4 YCbCr 模型及其转换

YCbCr 模型通过亮度和色差（color difference）等三个分量进行颜色描述。色度（chrominance）信息通过色差分量 Cb 和 Cr 表示，其中 Cb 表示蓝色分量与参考值之差；Cr 表示红色分量与参考值之差。

图 1.15 所示为 YCbCr 图像及其三个分量图像。图 1.15(a) 为 YCbCr 图像，图 1.15(b)、图 1.15(c) 和图 1.15(d) 分别为亮度、色差 Cb 和色差 Cr 等分量图像。

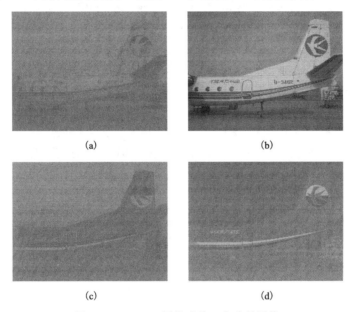

图 1.15　YCbCr 图像及其三个分量图像

RGB 图像的 R、G、B 分量与 YCbCr 图像的 Y、Cb、Cr 分量之间的转换关系可表示为

$$\begin{bmatrix} Y \\ Cb \\ Cr \end{bmatrix} = \begin{bmatrix} 65.481 & 128.553 & 24.966 \\ -37.797 & -74.203 & 112.000 \\ 112.000 & -93.786 & -18.214 \end{bmatrix} \begin{bmatrix} R \\ G \\ B \end{bmatrix} + \begin{bmatrix} 16 \\ 128 \\ 128 \end{bmatrix} \quad (1.8)$$

式中，R, G, B 取值范围为 [0, 1]。

MATLAB 采用 rgb2ycbcr 函数把 RGB 图像转换为 YCbCr 图像或采用 ycbcr2rgb 函数把 YCbCr 图像转换为 RGB 图像，其主要用法如下：

(1) YCBCR = rgb2ycbcr(RGB)，把 RGB 图像转换为 YCbCr 图像，其中 RGB 和 YCbCr 都是 M×N×3 矩阵。对于 8 位无符号整型图像，Y 的取值范围为 [16, 235]，Cb 和 Cr 的取值范围为 [16, 240]；对于 16 位无符号整型图像，Y 的取值范围为 [4112, 60395]，Cb 和 Cr 的

取值范围为[4112, 61689]；对于双精度浮点型图像，Y 的取值范围为[16/255, 235/255]，Cb 和 Cr 的取值范围为[16/255, 240/255]。

(2) RGB = ycbcr2rgb(YCBCR)，把 YCbCr 图像转换为 RGB 图像。

1.8　图像格式及其转换

图像格式是指图像文件的存储格式。利用 MATLAB 函数，首先读取一种格式的图像文件，然后再写为其他格式的图像文件，即可进行图像格式转换。

MATLAB 利用 imread 函数把图像从图像文件读到工作空间。MATLAB 在图像读入时支持以下图像格式：BMP(windows bitmap)、CUR(cursor file)、GIF(graphics interchange format)、HDF4(hierarchical data format)、ICO(icon file)、JPEG(joint photographic experts group)、JPEG 2000(joint photographic experts group 2000)、PBM(portable bitmap)、PCX(windows paintbrush)、PGM(portable graymap)、PNG(portable network graphics)、PPM(portable pixmap)、RAS(sun raster)、TIFF(tagged image file format)和 XWD(X window dump)等。

MATLAB 利用 imwrite 函数把图像从工作空间写到图像文件。MATLAB 在图像写出时支持以下图像格式：BMP、GIF、HDF、JPG 或 JPEG、JP2 或 JPX(即 JPEG 2000)、PBM、PCX、PGM、PNG、PNM(portable anymap)、PPM、RAS、TIF 或 TIFF 和 XWD 等。

第 2 章 图 像 运 算

图像运算(image operation)是指在空域对图像进行操作。常用图像运算主要包括几何运算、算术运算、卷积和相关等。

2.1 几 何 运 算

几何运算是指改变全部图像或部分图像的大小或形状,如图像插值、图像平移、图像缩放、图像旋转和图像剪切等。

2.1.1 图像插值

任何数字图像都只能给出整像素灰度,而不能给出亚像素灰度。然而,在图像处理中输出图像的整像素位置常常对应于输入图像的亚像素位置(即 4 个相邻整像素之间的位置),因此常常需要确定输入图像亚像素灰度。图像插值(image interpolation)可用于输入图像亚像素灰度确定,主要包括最近邻插值(nearest-neighbor interpolation)、双线性插值(bilinear interpolation)和双三次插值(bicubic interpolation)等。

1. 最近邻插值

最近邻插值是指亚像素灰度等于离它最近的整像素灰度,如图 2.1 所示。最近邻插值所得到的灰度分布在部分整像素位置出现间断。最近邻插值可表示为

$$f(x,y) = \begin{cases} f(0,0) & \left(0 \le x \le \frac{1}{2}; 0 \le y \le \frac{1}{2}\right) \\ f(1,0) & \left(\frac{1}{2} \le x \le 1; 0 \le y \le \frac{1}{2}\right) \\ f(0,1) & \left(0 \le x \le \frac{1}{2}; \frac{1}{2} \le y \le 1\right) \\ f(1,1) & \left(\frac{1}{2} \le x \le 1; \frac{1}{2} \le y \le 1\right) \end{cases} \tag{2.1}$$

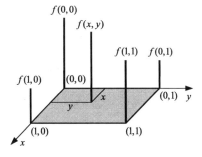

图 2.1 最近邻插值

式中，(x,y)为亚像素坐标，$f(x,y)$为亚像素灰度。

2. 双线性插值

双线性插值是指亚像素灰度等于离它最近的 4 个整像素灰度在相互垂直方向的线性加权平均，如图 2.2 所示。双线性插值所得到的灰度分布在整像素位置连续。双线性插值可表示为

$$f(x,y) = \begin{bmatrix} 1-x \\ x \end{bmatrix}^{\mathrm{T}} \begin{bmatrix} f(0,0) & f(0,1) \\ f(1,0) & f(1,1) \end{bmatrix} \begin{bmatrix} 1-y \\ y \end{bmatrix} \quad (0 \leqslant x \leqslant 1; 0 \leqslant y \leqslant 1) \quad (2.2)$$

式中，T 表示矩阵转置。

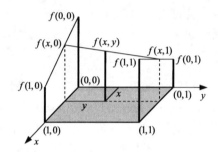

图 2.2 双线性插值

3. 双三次插值

双三次插值是指亚像素灰度等于离它最近的 16 个整像素灰度在相互垂直方向的三次加权平均，如图 2.3 所示。双三次插值所得到的灰度分布及其一阶导数在整像素位置连续。双三次插值可表示为

$$f(x,y) = \begin{bmatrix} S(1+x) \\ S(x) \\ S(1-x) \\ S(2-x) \end{bmatrix}^{\mathrm{T}} \begin{bmatrix} f(0,0) & f(0,1) & f(0,2) & f(0,3) \\ f(1,0) & f(1,1) & f(1,2) & f(1,3) \\ f(2,0) & f(2,1) & f(2,2) & f(2,3) \\ f(3,0) & f(3,1) & f(3,2) & f(3,3) \end{bmatrix} \begin{bmatrix} S(1+y) \\ S(y) \\ S(1-y) \\ S(2-y) \end{bmatrix} \quad (0 \leqslant x \leqslant 1; 0 \leqslant y \leqslant 1)$$

(2.3)

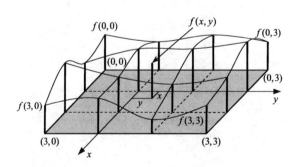

图 2.3 双三次插值

式中

$$S(t) = \begin{cases} 1-(a+3)|t|^2 +(a+2)|t|^3 & (|t| \leq 1) \\ -4a+8a|t|-5a|t|^2+a|t|^3 & (1 \leq |t| \leq 2) \end{cases} \quad (2.4)$$

式中，a 为常数，其取值通常等于 -0.5。当 $a = -0.5$ 时，则有

$$S(t) = \begin{cases} 1-2.5|t|^2+1.5|t|^3 & (|t| \leq 1) \\ 2-4|t|+2.5|t|^2-0.5|t|^3 & (1 \leq |t| \leq 2) \end{cases} \quad (2.5)$$

MATLAB 利用 interpn 函数进行一维、二维、三维和三维以上数组插值，其主要用法如下：

(1) B = interpn(X1, X2, …, Xn, A, Y1, Y2, …, Yn)，采用指定网格坐标，进行数组线性插值。

(2) B = interpn(A, Y1, Y2, …, Yn)，采用默认网格坐标 $(1,2,3,\cdots)$，进行数组线性插值。

(3) B = interpn(A, k)，网格进行 k 次划分后，完成数组线性插值。k 省略时，是指进行一次网格划分。

(4) B = interpn(…, method)，按照选项 method 指定的插值方法，进行数组插值。对二维及二维以上数组，method 选项包括线性插值('linear')、近邻插值('nearest')和三次插值('cubic')等。默认选项是线性插值。

图 2.4 所示为图像插值结果。插值前为 3×3 图像矩阵 $\begin{bmatrix} 0 & 0.5 & 0 \\ 0.5 & 1 & 0.5 \\ 0 & 0.5 & 0 \end{bmatrix}$，插值后为 750×1000 图像矩阵。图 2.4(a)为最近邻插值图像，图 2.4(b)为双线性插值图像，图 2.4(c)为双三次插值图像。

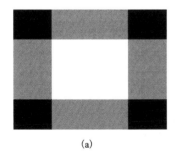 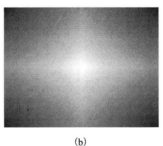 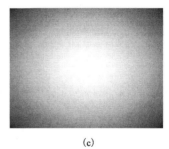

(a)   (b)   (c)

图 2.4 图像插值

2.1.2 图像平移

图像平移可表示为

$$\begin{bmatrix} x' \\ y' \\ 1 \end{bmatrix} = \begin{bmatrix} 1 & 0 & u \\ 0 & 1 & v \\ 0 & 0 & 1 \end{bmatrix} = \begin{bmatrix} x \\ y \\ 1 \end{bmatrix} \quad (2.6)$$

式中，(x,y) 和 (x',y') 分别为图像平移前后的像素坐标，u,v 为像素分别沿 x,y 方向的位移分量。

MATLAB 利用 imtranslate 函数平移图像，其主要用法如下：

(1) B = imtranslate (A, translation)，采用双线性插值方法平移图像，其中位移矢量由 translation 指定。

(2) B = imtranslate (A, translation, method)，采用指定插值方法平移图像，其中 method 是指近邻插值('nearest')、线性插值('bilinear')或三次插值('bicubic')，默认时是指线性插值。

图 2.5 所示为图像平移结果。图 2.5(a) 为原始真彩图像，图 2.5(b) 为平移后的真彩图像。

(a)　　　　　　　　　　(b)

图 2.5　图像平移

2.1.3　图像缩放

图像缩放可表示为

$$\begin{bmatrix} x' \\ y' \\ 1 \end{bmatrix} = \begin{bmatrix} m & 0 & 0 \\ 0 & n & 0 \\ 0 & 0 & 1 \end{bmatrix} = \begin{bmatrix} x \\ y \\ 1 \end{bmatrix} \tag{2.7}$$

式中，(x,y) 和 (x',y') 分别为图像缩放前后的像素坐标，m,n 分别为 x,y 方向的放大倍数。

MATLAB 利用 imresize 函数缩放图像，其主要用法如下：

(1) B = imresize (A, scale)，把图像 A 缩放 scale 倍。输入图像 A 可以是灰度图像、真彩图像或二值图像。如果 scale 在 0~1.0 之间取值，则缩小图像；如果 scale 大于 1.0，则放大图像。

(2) B = imresize (A, [numrows numcols])，把图像 A 缩放到 numrows 行和 numcols 列。

(3) [⋯] = imresize (⋯, method)，缩放图像，其中 method 是指最近邻插值('nearest')、双线性插值('bilinear')或双三次插值('bicubic')，默认时是指双三次插值。

图 2.6 所示为图像缩放结果。图 2.6(a) 为原始真彩图像，图 2.6(b) 为缩小后的真彩图像。

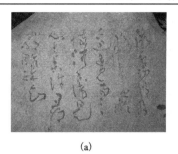 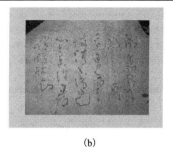

(a)　　　　　　　　　　　(b)

图 2.6　图像缩放

2.1.4　图像旋转

图像旋转可表示为

$$\begin{bmatrix} x' \\ y' \\ 1 \end{bmatrix} = \begin{bmatrix} \cos\theta & -\sin\theta & 0 \\ \sin\theta & \cos\theta & 0 \\ 0 & 0 & 1 \end{bmatrix} \begin{bmatrix} x \\ y \\ 1 \end{bmatrix} \tag{2.8}$$

式中，(x,y) 和 (x',y') 分别为图像旋转前后的像素坐标，θ 为图像逆时针旋转的角度。

MATLAB 利用 imrotate 函数旋转图像，其主要用法如下：

(1) B = imrotate(A, angle)，把图像 A 绕其中心旋转角度 angle。如果 angle 为正，则逆时针旋转；如果 angle 为负，则顺时针旋转。

(2) B = imrotate(A, angle, method)，旋转图像，其中 method 可设定为最近邻插值('nearest')、双线性插值('bilinear')或双三次插值('bicubic')，默认时是指最近邻插值。

(3) B = imrotate(A, angle, method, bbox)，该函数旋转图像，其中 bbox 可设定为'crop'(输出图像与输入图像具有相同尺寸)或'loose'(输出图像完全包含输入图像，因此比输入图像大)，默认时是指'loose'。

图 2.7 所示为图像旋转结果。图 2.7(a)为原始真彩图像，图 2.7(b)为旋转后的真彩图像，其中 bbox 设定为'crop'。

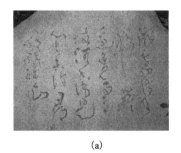 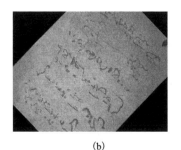

(a)　　　　　　　　　　　(b)

图 2.7　图像旋转

2.1.5　图像剪切

图像剪切可表示为

$$B(x,y) \subset A(x,y) \tag{2.9}$$

式中，$\subset$ 表示包含于，A 和 B 分别表示剪切前后的图像。

MATLAB 利用 imcrop 函数剪切图像，其主要用法如下：

(1) A = imcrop(I, rect)，剪切图像，其中 rect 表示[xmin ymin width height]，用以确定剪切区域的位置和尺寸。

(2) A = imcrop(X, map, rect)，剪切索引图像。

(3) [A rect] = imcrop(…)，剪切图像，并把剪切区域的位置和尺寸返回到 rect 中。

图 2.8 所示为图像剪切结果。图 2.8(a) 为原始真彩图像，图 2.8(b) 为剪切后的真彩图像。

图 2.8 图像剪切

2.2 算术运算

2.2.1 图像相加

两幅图像 A 和 B 相加定义为

$$C(x,y) = A(x,y) + B(x,y) \tag{2.10}$$

式中，A 和 B 具有相同行和相同列。

MATLAB 利用 imadd 函数进行图像相加，其主要用法如下：

Z = imadd(X, Y)，数组 X 和 Y 相加后得到数组 Z，其中 X 和 Y 必须具有相同尺寸和相同数据类型。除非指定 Z 的数据类型，否则 Z 将具有与 X(或 Y)相同的数据类型。如果 X 和 Y 是逻辑型，则 Z 是双精度型。

如果 X 和 Y 是整型数组，那么 Z 中超出整型范围的数据将被截去超出部分。例如，当两个 8 位无符号整型数组 $X = \begin{bmatrix} 0 & 100 \\ 200 & 255 \end{bmatrix}$ 和 $Y = \begin{bmatrix} 100 & 150 \\ 200 & 255 \end{bmatrix}$ 相加时，由于 8 位无符号整型数据只能处于[0, 255]范围，即相加后的数据超过 255 时则会被截断为 255，因此 X 和 Y 相加之后的数组 $Z = \begin{bmatrix} 100 & 250 \\ 255 & 255 \end{bmatrix}$。上述数组相加的 M 文件为

$$X = \text{uint8}([0\ 100;\ 200\ 255]);$$
$$Y = \text{uint8}([100\ 150;\ 200\ 255]);$$
$$Z = \text{imadd}(X, Y)$$

运行结果为

$$Z =$$
$$100\ \ 250$$
$$255\ \ 255$$

数组类型为

Name	Size	Bytes	Class	Attributes
X	2×2	4	uint8	
Y	2×2	4	uint8	
Z	2×2	4	uint8	

为了避免数据被截断，可以指定输出数组 Z 的数据类型，如指定 Z 为 16 位无符号整型，则 X 和 Y 相加之后的数组 $Z = \begin{bmatrix} 100 & 250 \\ 400 & 510 \end{bmatrix}$。上述数组相加的 M 文件为

$$X = \text{uint8}([0\ 100;\ 200\ 255]);$$
$$Y = \text{uint8}([100\ 150;\ 200\ 255]);$$
$$Z = \text{imadd}(X, Y, \text{'uint16'})$$

运行结果为

$$Z =$$
$$100\ \ 250$$
$$400\ \ 510$$

数组类型为

Name	Size	Bytes	Class	Attributes
X	2×2	4	uint8	
Y	2×2	4	uint8	
Z	2×2	8	uint16	

如果 X 和 Y 是逻辑型，当他们相加时，则 Z 是双精度型。例如，两个逻辑型数组 $X = \begin{bmatrix} 0 & 0 \\ 1 & 1 \end{bmatrix}$ 和 $Y = \begin{bmatrix} 0 & 1 \\ 0 & 1 \end{bmatrix}$ 相加后的数组 $Z = \begin{bmatrix} 0 & 1 \\ 1 & 2 \end{bmatrix}$。上述数组相加的 M 文件为

$$X = \text{logical}([0\ 0;\ 1\ 1]);$$
$$Y = \text{logical}([0\ 1;\ 0\ 1]);$$
$$Z = \text{imadd}(X, Y)$$

运行结果为

$$Z =$$
$$0\ \ 1$$
$$1\ \ 2$$

数组类型为

Name	Size	Bytes	Class	Attributes
X	2×2	4	logical	
Y	2×2	4	logical	
Z	2×2	32	double	

图 2.9 所示为图像相加结果。图 2.9(a)和图 2.9(b)为原始真彩图像，图 2.9(c)为原始真彩图像相加后的真彩图像。

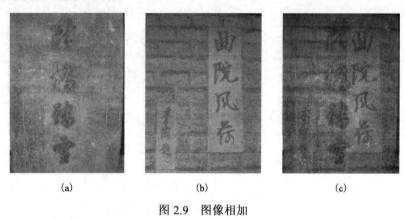

(a)   (b)   (c)

图 2.9　图像相加

2.2.2　图像相减

两幅图像 A 和 B 相减定义为

$$C(x,y) = A(x,y) - B(x,y) \tag{2.11}$$

式中，A 和 B 具有相同行和相同列。

MATLAB 利用 imsubtract 函数进行图像相减，其主要用法如下：

Z = imsubtract(X, Y)，数组 X 和 Y 相减后得到数组 Z，其中 X 和 Y 必须具有相同尺寸和相同数据类型。除非 X(或 Y)是逻辑型时 Z 是双精度型，否则 Z 具有与 X(或 Y)相同的数据类型。

如果 X 是整型数组，那么超出整型范围的数据将被截断。例如，当两个 8 位无符号整型数组 $X = \begin{bmatrix} 0 & 100 \\ 200 & 255 \end{bmatrix}$ 和 $Y = \begin{bmatrix} 100 & 150 \\ 200 & 255 \end{bmatrix}$ 相减时，由于 8 位无符号整型数据只能处于[0, 255]范围，即相减后的数据为负值时则会被截断为 0，因此 X 和 Y 相减之后的数组 $Z = \begin{bmatrix} 0 & 0 \\ 0 & 0 \end{bmatrix}$。上述数组相减的 M 文件为

    X = uint8([0 100; 200 255]);
    Y = uint8([100 150; 200 255]);
    Z = imsubtract(X, Y)

运行结果为

$$Z = \begin{matrix} 0 & 0 \\ 0 & 0 \end{matrix}$$

数组类型为

Name	Size	Bytes	Class	Attributes
X	2×2	4	uint8	
Y	2×2	4	uint8	
Z	2×2	4	uint8	

为了避免数据被截断,可以先把数组 X 和 Y 的数据类型由 8 位无符号整型转换为 16 位整型,则 X 和 Y 相减之后的数组 $Z = \begin{bmatrix} -100 & -50 \\ 0 & 0 \end{bmatrix}$。上述数组相减的 M 文件为

X = int16(uint8([0 100; 200 255]));
Y = int16(uint8([100 150; 200 255]));
Z = imadd(X, Y)

运行结果为

$$Z = \begin{matrix} -100 & -50 \\ 0 & 0 \end{matrix}$$

数组类型为

Name	Size	Bytes	Class	Attributes
X	2×2	8	int16	
Y	2×2	8	int16	
Z	2×2	8	int16	

图 2.10 所示为图像相减结果。图 2.10(a)为原始真彩图像,图 2.10(b)为原始真彩图像亮度分量的背景图像,图 2.10(c)为原始真彩图像与亮度背景图像相减后的真彩图像。

(a)　　　　　　　　　　(b)　　　　　　　　　　(c)

图 2.10　图像相减

2.2.3　绝对差值

两幅图像 A 和 B 绝对差值定义为

$$C(x, y) = |A(x, y) - B(x, y)| \qquad (2.12)$$

式中，$|\cdots|$ 表示求绝对值，A 和 B 具有相同行和相同列。

MATLAB 利用 imabsdiff 函数计算图像绝对差值，其主要用法如下：

Z = imabsdiff(X, Y)，数组 X 和 Y 相减，绝对差值返回到数组 Z。如果 X 和 Y 是整型数组，那么超出整型范围的数据将被截断。如果 X 和 Y 是双精度型数组，那么 imabsdiff(X, Y) 等效于 abs(X − Y)。如果 X 和 Y 是逻辑型数组，那么 imabsdiff(X, Y) 等效于 XOR(X − Y)。

图 2.11 所示为绝对差值计算结果。图 2.11(a) 和图 2.11(b) 为原始真彩图像，图 2.11(c) 为绝对差值真彩图像。

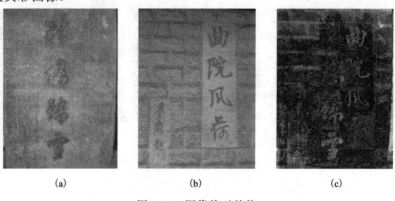

图 2.11　图像绝对差值

2.2.4　图像相乘

两幅图像 A 和 B 相乘定义为

$$C(x, y) = A(x, y) \times B(x, y) \tag{2.13}$$

式中，A 和 B 具有相同行和相同列。

MATLAB 利用 immultiply 函数进行图像相乘，其主要用法如下：

Z = immultiply(X, Y)，数组 X 和 Y 的元素相乘后得到数组 Z。如果 X 和 Y 是具有相同尺寸和相同数据类型，那么 Z 具有与 X(或 Y)相同的尺寸和数据类型。如果 X(或 Y)是数值型而 Y(或 X)是逻辑型，那么 Z 具有与 X(或 Y)相同的尺寸和数据类型。如果 X 和 Y 是整型数组，那么超出整型范围的数值将被截断。

图 2.12 所示为图像相乘结果。图 2.12(a) 和图 2.12(b) 为原始真彩图像，图 2.12(c) 为原始真彩图像相乘后的真彩图像。

 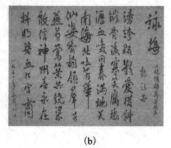 

图 2.12　图像相乘

2.2.5 图像相除

两幅图像 A 和 B 相除定义为

$$B(x,y) = A(x,y) \div B(x,y) \tag{2.14}$$

式中，A 和 B 具有相同行和相同列。

MATLAB 利用 imdivide 函数进行图像相除，其主要用法如下：

Z = imdivide(X, Y)，数组 X 和 Y 的元素相除后得到数组 Z，其中 X 和 Y 具有相同大小和相同数据类型。Z 具有与 X(或 Y)相同的尺寸和数据类型。如果 X 和 Y 是整型数组，那么超出整型范围的数据将被截断，同时小数值将四舍五入到最近整数。

图 2.13 所示为图像相除结果。图 2.13(a)和图 2.13(b)为原始真彩图像，图 2.13(c)为原始真彩图像相除后的真彩图像。

  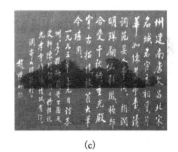

(a)　　　　　　　　　　　(b)　　　　　　　　　　　(c)

图 2.13　图像相除

2.2.6 图像线性组合

图像 $A_1, A_2, A_3, \cdots, A_n$ 的线性组合定义为

$$B(x,y) = k_1 A_1(x,y) + k_2 A_2(x,y) + k_3 A_3(x,y) + \cdots + k_n A_n(x,y) \tag{2.15}$$

式中，k 表示每幅图像所占的权重。需要注意的是，只有具有相同行和相同列的图像才能进行线性组合。

MATLAB 利用 imlincomb 函数进行图像线性组合，其主要用法如下：

(1) Z = imlincomb(K1, A1, K2, A2, ···, Kn, An)，计算线性组合数组 K1×A1+K2×A2 +···+Kn×An，其中 K1, K2, ···, Kn 是双精度型实数、A1, A2, ···, An 是具有相同尺寸和数据类型的数组。数组 Z 的尺寸和数据类型与 A1 相同。

(2) Z = imlincomb(K1, A1, K2, A2, ···, Kn, An, K)，计算计算线性组合数组，其中 K 是双精度型实数。

(3) Z = imlincomb(···, output_class)，计算线性组合数组，其中 output_class 表示数组 Z 的数据类型。

如果 Z 是整型数组，那么超出整型范围的数值将被截断。

2.3 卷积与相关运算

2.3.1 图像卷积

两幅图像 A 和 B 的卷积定义为

$$C(x,y) = A(x,y) * B(x,y) \quad (2.16)$$

式中，$*$ 表示图像卷积。设 A 和 B 的尺寸分别为 $M \times N$ 和 $P \times Q$，则上式也可表示为

$$C(x,y) = \sum_{m=1}^{M}\sum_{n=1}^{N} A(m,n)B(x-m+1, y-n+1) \quad \begin{cases} x=1,2,\cdots,M+P-1 \\ y=1,2,\cdots,N+Q-1 \end{cases} \quad (2.17)$$

式中，图像 C 的尺寸为 $(M+P-1) \times (N+Q-1)$。

MATLAB 利用 conv2 函数进行图像二维卷积运算，其主要用法如下：

(1) C = conv2(A, B)，计算数组 A 和 B 的二维卷积。如果其中一个数组为二维有限脉冲响应(Finite Impulse Response, FIR)滤波器，则卷积运算即为对另一数组进行二维滤波。如果数组 A 和 B 的尺寸分别为 $M \times N$ 和 $P \times Q$，则 C 的尺寸为 $\max\{M+P-1, M, P\} \times \max\{N+Q-1, N, Q\}$，其中 $\max\{\cdots\}$ 表示取最大值。

(2) C = conv2(A, B, shape)，返回数组 X 和 Y 的二维卷积，区域大小由 shape 指定。当 shape 指定为 full 时，返回整个二维卷积区域(缺省情况)；当 shape 指定为 same 时，返回与 A 尺寸相同的中心部分二维卷积区域；当 shape 指定为 valid 时，仅返回 A 边缘不用 0 填充而得到的二维卷积。

2.3.2 图像相关

两幅图像 A 和 B 的相关定义为

$$C(x,y) = A(x,y) \circ B(x,y) \quad (2.18)$$

式中，$\circ$ 表示图像相关。设 A 和 B 的尺寸分别为 $M \times N$ 和 $P \times Q$，则上式也可表示为

$$C(x,y) = \sum_{m=1}^{M}\sum_{n=1}^{N} A(m,n)B*(m-x+P, n-y+Q) \quad \begin{cases} x=1,2,\cdots,M+P-1 \\ y=1,2,\cdots,N+Q-1 \end{cases} \quad (2.19)$$

式中，图像 C 的尺寸为 $(M+P-1) \times (N+Q-1)$。

MATLAB 利用 xcorr2 函数进行图像二维互相关运算，其主要用法如下：

(1) C = xcorr2(A, B)，在不进行缩放的情况下，计算数组 A 和 B 的二维互相关。

(2) C = xcorr2(A)，计算数组 A 的二维自相关。xcorr2(A) 等同于 xcorr2(A, A)。

2.3.3 卷积和相关应用

图 2.14 所示为基于卷积和相关的图像平滑(即空域低通滤波)结果。图 2.14(a)为原

始灰度图像,图 2.14(b)为原图与 $h = \begin{bmatrix} 1 & 1 & \cdots & 1 \\ 1 & 1 & \cdots & 1 \\ \vdots & \vdots & & \vdots \\ 1 & 1 & \cdots & 1 \end{bmatrix}_{15 \times 15}$ 进行卷积运算后的图像,图 2.14(c)

为原图与 $h = \begin{bmatrix} 1 & 1 & \cdots & 1 \\ 1 & 1 & \cdots & 1 \\ \vdots & \vdots & & \vdots \\ 1 & 1 & \cdots & 1 \end{bmatrix}_{15 \times 15}$ 进行相关运算后的图像。

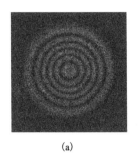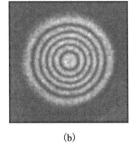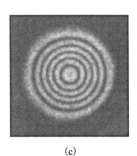

(a) (b) (c)

图 2.14 卷积、相关平滑结果

图 2.15 所示为基于卷积和相关的图像锐化(即空域高通滤波)结果。图 2.15(a)为原始灰度图像,图 2.15(b)为原图与 $h = \begin{bmatrix} -1 & -1 & -1 \\ 0 & 0 & 0 \\ 1 & 1 & 1 \end{bmatrix}$ 进行卷积运算后的图像,图 2.15(c)

为原图与 $h = \begin{bmatrix} -1 & -1 & -1 \\ 0 & 0 & 0 \\ 1 & 1 & 1 \end{bmatrix}$ 进行相关运算后的图像。

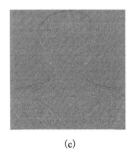

(a) (b) (c)

图 2.15 卷积、相关的图像锐化结果

图 2.16 所示为基于相关的图像特征定位结果。图 2.16(a)为原始二值图像(2700×3600),图 2.16(b)为模板(385×373),图 2.16(c)为原图与模板进行相关后的灰度图像,图 2.16(d)为显示最大灰度所在位置(即与模板最佳匹配的图像区域中心在原图中的位置)的二值图像。

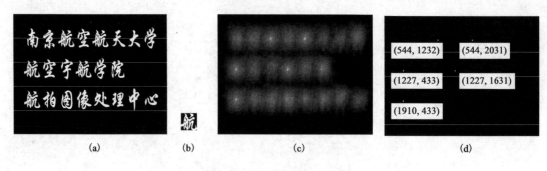

图 2.16　图像特征定位结果

第3章 图像变换

图像变换(image transformation)是指将图像从空域变换到频域,在频域对图像进行处理,再把处理结果从频域反变换到空域,进而得到所需要的图像。常用图像变换主要包括傅里叶变换和余弦变换等。

3.1 傅里叶变换

傅里叶变换(Fourier transform)描述了离散信号的空域表示与频域表示之间的关系,是信号处理的有效工具之一,对频谱分析、卷积与相关运算、滤波处理、功率谱分析和传递函数建模等的快速计算起到了关键作用。利用傅里叶变换可解决很多图像处理问题,因而傅里叶变换在图像处理领域具有广泛应用。

3.1.1 连续傅里叶变换

1. 一维连续傅里叶变换

设函数 $f(x)$ 满足:①在无限空间绝对可积;②具有有限个间断点和有限个极值点;③没有无穷大间断点,则该函数存在傅里叶变换。函数 $f(x)$ 的一维连续傅里叶变换定义为

$$F(u) = \int_{-\infty}^{\infty} f(x)\exp(-\mathrm{i}2\pi ux)\mathrm{d}x \tag{3.1}$$

式中,$\mathrm{i}=\sqrt{-1}$ 为虚数单位,u 为连续频率变量。

一维连续傅里叶反变换定义为

$$f(x) = \int_{-\infty}^{\infty} F(u)\exp(\mathrm{i}2\pi xu)\mathrm{d}u \tag{3.2}$$

2. 二维连续傅里叶变换

函数 $f(x,y)$ 的二维连续傅里叶变换定义为

$$F(u,v) = \int_{-\infty}^{\infty}\int_{-\infty}^{\infty} f(x,y)\exp[-\mathrm{i}2\pi(ux+vy)]\mathrm{d}x\mathrm{d}y \tag{3.3}$$

式中,u,v 分别为沿 x,y 方向的连续频率变量。

二维连续傅里叶反变换定义为

$$f(x,y) = \int_{-\infty}^{\infty}\int_{-\infty}^{\infty} F(u,v)\exp[\mathrm{i}2\pi(ux+vy)]\mathrm{d}u\mathrm{d}v \tag{3.4}$$

3.1.2 傅里叶变换性质

1. 幅值和相位

函数 $f(x,y)$ 的傅里叶变换 $F(u,v)$ 通常为复函数,那么 $F(u,v)$ 可表示为

$$F(u,v) = R(u,v) + \mathrm{i}I(u,v) \tag{3.5}$$

式中,i 为虚数单位,$R(u,v)$ 和 $I(u,v)$ 分别为 $F(u,v)$ 的实部和虚部。通过实部和虚部,则频谱幅值和相位可分别表示为

$$|F(u,v)| = \sqrt{R^2(u,v) + I^2(u,v)}$$
$$\varphi(u,v) = \arctan[I(u,v)/R(u,v)] \tag{3.6}$$

值得注意的是,在数字图像处理中,频谱幅值分布通常不是由 $|F(u,v)|$ 直接显示,而是由 $\ln(1+|F(u,v)|)$ 进行显示。

图 3.1 所示为频谱幅值和相位分布。图 3.1(a) 为原始灰度图像,图 3.1(b) 为频谱幅值图像,图 3.1(c) 为频谱相位图像。

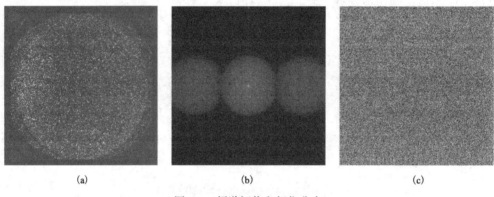

图 3.1　频谱幅值和相位分布

2. 共轭对称性

设 $f(x,y)$ 为实函数,则其傅里叶变换关于原点共轭对称,即

$$F(u,v) = F^*(-u,-v) \tag{3.7}$$

式中,* 表示复共轭。对频谱幅值,有

$$|F(u,v)| = |F(-u,-v)| \tag{3.8}$$

即频谱幅值关于原点对称,如图 3.1(b) 所示。

3. 旋转不变性

若 $\mathrm{FT}\{f(x,y)\} = F(u,v)$,则有

$$\mathrm{FT}\{f(r,\theta+\alpha)\} = F(\rho,\varphi+\alpha) \tag{3.9}$$

式中,$\mathrm{FT}\{\cdots\}$ 表示傅里叶变换,α 为空域旋转角度,(r,θ) 和 (ρ,φ) 分别为空域和频域的

极坐标，即 $x = r\cos\theta, y = r\sin\theta$ 和 $u = \rho\cos\varphi, v = \rho\sin\varphi$。上式表明，当 $f(x,y)$ 在空域发生旋转，其傅里叶变换 $F(u,v)$ 在频域将会出现同等程度的旋转，即转向相同、转角相等。

图 3.2 所示为空域、频域旋转结果。图 3.2(a) 为空域旋转图像，图 3.2(b) 为频域幅值旋转图像，图 3.2(c) 为频域相位旋转图像。

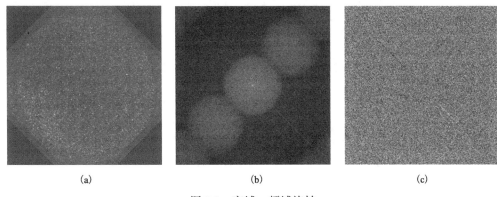

图 3.2　空域、频域旋转

4. 线性定理

若 $\text{FT}\{f(x,y)\} = F(u,v)$ 和 $\text{FT}\{g(x,y)\} = G(u,v)$，则有

$$\text{FT}\{\alpha f(x,y) + \beta g(x,y)\} = \alpha F(u,v) + \beta G(u,v) \tag{3.10}$$

式中，α 和 β 为常数。上式表明，函数之和的傅里叶变换等于它们傅里叶变换之和。

5. 相似定理

若 $\text{FT}\{f(x,y)\} = F(u,v)$，则

$$\text{FT}\{f(\alpha x, \beta y)\} = \frac{1}{|\alpha\beta|} F\left(\frac{u}{\alpha}, \frac{v}{\beta}\right) \quad (\alpha, \beta \neq 0) \tag{3.11}$$

式中，α 和 β 为常数。上式表明，空域中坐标 (x,y) 的伸展将导致频域中坐标 (u,v) 的收缩和频谱幅度的变化。

6. 相移定理

若 $\text{FT}\{f(x,y)\} = F(u,v)$，则

$$\text{FT}\{f(x-a, y-b)\} = F(u,v)\exp[-\text{i}2\pi(ua+vb)] \tag{3.12}$$

式中，a 和 b 为常数。上式表明，函数 $f(x,y)$ 在空域位移，将引起频谱在频域相移，但频谱幅值保持不变，即

$$|F(u,v)\exp[-\text{i}2\pi(ua+vb)]| = |F(u,v)| \tag{3.13}$$

图 3.3 所示为空域位移、频域相移结果。图 3.3(a) 为空域位移图像，图 3.3(b) 为频域幅值图像，图 3.3(c) 为频域相位图像。

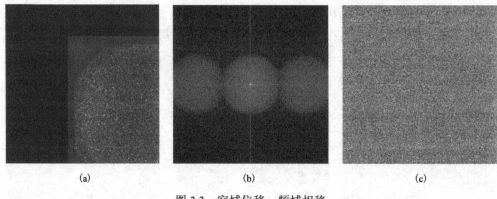

图 3.3 空域位移、频域相移

7. 巴塞伐尔(Parseval)定理

若 $FT\{f(x,y)\} = F(u,v)$，则

$$\int_{-\infty}^{\infty}\int_{-\infty}^{\infty} |f(x,y)|^2 \, dxdy = \int_{-\infty}^{\infty}\int_{-\infty}^{\infty} |F(u,v)|^2 \, dxdy \tag{3.14}$$

上式表示能量守恒。

8. 卷积定理

函数 $f(x,y)$ 和 $g(x,y)$ 的卷积定义为

$$f(x,y) * g(x,y) = \int_{-\infty}^{\infty}\int_{-\infty}^{\infty} f(\alpha,\beta)g(x-\alpha, y-\beta)d\alpha d\beta \tag{3.15}$$

式中，*表示卷积。设 $FT\{f(x,y)\} = F(u,v)$ 和 $FT\{g(x,y)\} = G(u,v)$，则有

$$FT\{f(x,y) * g(x,y)\} = F(u,v)G(u,v) \tag{3.16}$$

上式表明，空域中两个函数卷积的傅里叶变换等于频域中每个函数傅里叶变换的乘积。

9. 相关定理

函数 $f(x,y)$ 和 $g(x,y)$ 的互相关定义为

$$f(x,y) \circ g(x,y) = \int_{-\infty}^{\infty}\int_{-\infty}^{\infty} f(\alpha,\beta)g^*(\alpha-x, \beta-y)d\alpha d\beta \tag{3.17}$$

式中，。表示相关，上标*表示复共轭。

当 $g(x,y) = f(x,y)$，则有

$$FT\{f(x,y) \circ f(x,y)\} = F(u,v)F^*(u,v) = |F(u,v)|^2 \tag{3.18}$$

上式即为自相关定理。

3.1.3 离散傅里叶变换

1. 一维离散傅里叶变换

设 $f(x)$ 是在时域上等间隔采样得到 M 点离散信号，其中 x 是离散实变量，即 $x = 0, 1, \cdots, M-1$，则一维离散傅里叶变换定义为

$$F(u) = \sum_{x=0}^{M-1} f(x) \exp\left(-i2\pi \frac{xu}{M}\right) \quad (u = 0, 1, \cdots, M-1) \tag{3.19}$$

式中，u 为离散频率变量；$\exp\left(-i2\pi \dfrac{xu}{M}\right)$ 为变换核。

一维离散傅里叶反变换定义为

$$f(x) = \frac{1}{M} \sum_{u=0}^{M-1} F(u) \exp\left(i2\pi \frac{xu}{M}\right) \quad (x = 0, 1, \cdots, M-1) \tag{3.20}$$

式中，$\exp\left(i2\pi \dfrac{xu}{M}\right)$ 为反变换核。

2. 二维离散傅里叶变换

设 $f(x, y)$ 是在空域上等间隔采样得到 $M \times N$ 点二维离散信号，其中 x, y 是离散实变量，则二维离散傅里叶变换定义为

$$F(u, v) = \sum_{x=0}^{M-1} \sum_{y=0}^{N-1} f(x, y) \exp\left[-i2\pi\left(\frac{xu}{M} + \frac{yv}{N}\right)\right] \quad \begin{pmatrix} u = 0, 1, \cdots, M-1 \\ v = 0, 1, \cdots, N-1 \end{pmatrix} \tag{3.21}$$

式中，u, v 为离散频率变量。

二维离散傅里叶反变换定义为

$$f(x, y) = \frac{1}{MN} \sum_{u=0}^{M-1} \sum_{v=0}^{N-1} F(u, v) \exp\left[i2\pi\left(\frac{xu}{M} + \frac{yv}{N}\right)\right] \quad \begin{pmatrix} x = 0, 1, \cdots, M-1 \\ y = 0, 1, \cdots, N-1 \end{pmatrix} \tag{3.22}$$

设空域和频域的采样间隔分别由 $\Delta x, \Delta y$ 和 $\Delta u, \Delta v$ 表示，则它们满足

$$\Delta u = \frac{1}{M \Delta x}, \quad \Delta v = \frac{1}{N \Delta y} \tag{3.23}$$

式中，M, N 分别为 x, y 方向的空域采样点数。

3.1.4 快速傅里叶变换

在数字图像处理中，当图像阵列较大时，直接采用离散傅里叶变换往往具有很大的计算量。为了减小计算量，人们提出了快速傅里叶变换(fast Fourier transform)。快速傅里叶变换就是离散傅里叶变换的快速算法，它将离散傅里叶变换的乘法运算转变为加(减)法运算。快速傅里叶变换的提出为离散傅里叶变换的广泛应用奠定了基础。

1. 一维快速傅里叶变换

设 W_M^{xu} 表示变换核，即

$$W_M^{xu} = \exp\left(-i2\pi \frac{xu}{M}\right) \tag{3.24}$$

则一维离散傅里叶变换可表示为

$$F(u) = \sum_{x=0}^{M-1} f(x) W_M^{xu} \quad (u = 0, 1, \cdots, M-1) \tag{3.25}$$

上式表明,每计算一个 $F(u)$,需要进行 M 次乘法和 $(M-1)$ 次加法。对 M 个采样点,则要进行 M^2 次乘法和 $M(M-1)$ 次加法。当 M 很大时,计算量将非常大。如果将偶数项和奇数项分开,则式(3.25)可表示为

$$F(u) = \sum_{x=0}^{M/2-1} f(2x) W_M^{2xu} + \sum_{x=0}^{M/2-1} f(2x+1) W_M^{(2x+1)u} \quad (u = 0, 1, \cdots, M/2-1) \tag{3.26}$$

利用 $W_M^{2xu} = W_{M/2}^{xu}$,得

$$F(u) = \sum_{x=0}^{M/2-1} f(2x) W_{M/2}^{xu} + W_M^u \sum_{x=0}^{M/2-1} f(2x+1) W_{M/2}^{xu} \quad (u = 0, 1, \cdots, M/2-1) \tag{3.27}$$

设 $F_e(u) = \sum_{x=0}^{M/2-1} f(2x) W_{M/2}^{xu}$ 和 $F_o(u) = \sum_{x=0}^{M/2-1} f(2x+1) W_{M/2}^{xu}$,则上式可表示为

$$F(u) = F_e(u) + W_M^u F_o(u) \quad (u = 0, 1, \cdots, M/2-1) \tag{3.28}$$

式中,$F_e(u)$ 和 $F_o(u)$ 分别表示偶数项和奇数项;u 的取值范围为 $(0, 1, \cdots, M/2-1)$,而不是 $(0, 1, \cdots, M-1)$,因此还需要考虑取值范围 $(M/2, M/2+1, \cdots, M-1)$。

在 $(M/2, M/2+1, \cdots, M-1)$ 范围的 $F(u+M/2)$ 可表示为

$$F(u+M/2) = \sum_{x=0}^{M/2-1} f(2x) W_M^{2x(u+M/2)} + \sum_{x=0}^{M/2-1} f(2x+1) W_M^{(2x+1)(u+M/2)} \quad (u = 0, 1, \cdots, M/2-1) \tag{3.29}$$

利用 $W_M^{2xu} = W_{M/2}^{xu}$、$W_M^{xM} = 1$ 和 $W_M^{M/2} = -1$,得

$$F(u+M/2) = \sum_{x=0}^{M/2-1} f(2x) W_{M/2}^{xu} - W_M^u \sum_{x=0}^{M/2-1} f(2x+1) W_{M/2}^{xu}$$
$$= F_e(u) - W_M^u F_o(u) \quad (u = 0, 1, \cdots, M/2-1) \tag{3.30}$$

上述表明,快速傅里叶变换首先将原函数分为偶数项和奇数项,然后不断将偶数项和奇数项相加(减),进而得到所需要的结果。快速傅里叶变换的步骤是:第 1 步,将 1 个 M 点的离散傅里叶变换转化为 2 个 $M/2$ 点的离散傅里叶变换;第 2 步,将 2 个 $M/2$ 点的离散傅里叶变换转化为 4 个 $M/4$ 点的离散傅里叶变换;依次类推。

2. 二维快速傅里叶变换

二维离散傅里叶变换也可表示为

$$F(u,v) = \sum_{x=0}^{M-1} \sum_{y=0}^{N-1} f(x,y) W_M^{xu} W_N^{yv} \quad \begin{pmatrix} u = 0, 1, \cdots, M-1 \\ v = 0, 1, \cdots, N-1 \end{pmatrix} \tag{3.31}$$

式中,$W_M^{xu} = \exp\left(-\mathrm{i}2\pi \dfrac{xu}{M}\right)$;$W_N^{yv} = \exp\left(-\mathrm{i}2\pi \dfrac{yv}{N}\right)$。另外,上式还可分别表示为

$$F(u,v) = \sum_{x=0}^{M-1} \left[\sum_{y=0}^{N-1} f(x,y) W_N^{yv} \right] W_M^{xu} \quad \begin{pmatrix} u = 0, 1, \cdots, M-1 \\ v = 0, 1, \cdots, N-1 \end{pmatrix} \tag{3.32}$$

或

$$F(u,v) = \sum_{y=0}^{N-1}\left[\sum_{x=0}^{M-1} f(x,y) W_M^{xu}\right] W_N^{yv} \quad \begin{pmatrix} u=0,1,\cdots,M-1 \\ v=0,1,\cdots,N-1 \end{pmatrix} \qquad (3.33)$$

上式表明，二维快速傅里叶变换可以转化为两次一维快速傅里叶变换，即可先对图像的各列(或行)取快速傅里叶变换，然后再对各行(或列)取快速傅里叶变换。经过相互垂直方向的两次一维快速傅里叶变换，即可实现二维快速傅里叶变换。

3.1.5 离散傅里叶变换算法

1. 一维离散傅里叶变换

MATLAB 利用 fft 函数通过快速算法实现一维离散傅里叶变换，其主要用法如下：

(1) Y = fft(X)，通过快速算法进行一维离散傅里叶变换，其中 X 和 Y 具有相同尺寸。如果 X 是矩阵，那么 fft 函数对矩阵的每列分别进行傅里叶变换。如果 X 是多维数组，那么 fft 函数沿第一个非单个元素维方向进行傅里叶变换。例如：

$$X = \begin{bmatrix} 1 & 2 \end{bmatrix}, \text{则 } Y = \begin{bmatrix} 3 & -1 \end{bmatrix}$$

$$X = \begin{bmatrix} 1 & 2 \\ 3 & 4 \end{bmatrix}, \text{则 } Y = \begin{bmatrix} 4 & 6 \\ -2 & -2 \end{bmatrix}$$

$$X(:,:,1) = \begin{bmatrix} 1 & 2 \end{bmatrix}, \quad X(:,:,2) = \begin{bmatrix} 3 & 4 \end{bmatrix}, \text{则 } Y(:,:,1) = \begin{bmatrix} 3 & -1 \end{bmatrix}, \quad Y(:,:,2) = \begin{bmatrix} 7 & -1 \end{bmatrix}$$

图 3.4 所示为一维离散傅里叶变换结果。图 3.4(a) 为原始灰度图像，图 3.4(b) 为原始灰度图像经过一维傅里叶变换(列变换)后的频谱图像。

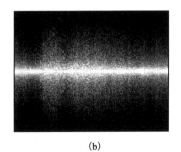

(a) （b）

图 3.4 一维离散傅里叶变换

(2) Y = fft(X, n)，通过快速算法进行一维离散傅里叶变换，其中 n 为指定的变换点数。如果 X 沿第一个非单个元素维方向的长度小于 n，则 X 沿该方向的尾部用 0 填充，使该方向的长度等于 n。如果 X 沿第一个非单个元素维方向的长度大于 n，则 X 沿该方向的数据被截断，使该方向的长度等于 n。

(3) Y = fft(X, [], dim) 和 Y = fft(X, n, dim)，沿第 dim 维方向进行离散傅里叶变换。

图 3.5 所示为一维离散傅里叶变换结果。图 3.5(a) 为原始灰度图像，图 3.5(b) 为原始灰度图像经过一维傅里叶变换(沿第二维方向变换)后的频谱图像。

 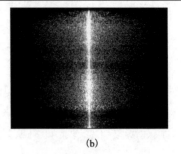

(a)   (b)

图 3.5   一维离散傅里叶变换

MATLAB 利用 ifft 函数通过快速算法实现一维离散傅里叶反变换，其主要用法如下：

(1) Y = ifft(X)，通过快速算法进行一维离散傅里叶反变换，其中 X 和 Y 具有相同尺寸。如果 X 是矩阵，那么 ifft 函数对矩阵每列分别进行傅里叶反变换。如果 X 是多维数组，那么 ifft 函数沿第一个非单个元素维方向进行傅里叶变换。例如：

$$X = \begin{bmatrix} 3 & -1 \end{bmatrix}, \text{则 } Y = \begin{bmatrix} 1 & 2 \end{bmatrix}$$

$$X = \begin{bmatrix} 4 & 6 \\ -2 & -2 \end{bmatrix}, \text{则 } Y = \begin{bmatrix} 1 & 2 \\ 3 & 4 \end{bmatrix}$$

$X(:,:,1) = \begin{bmatrix} 3 & -1 \end{bmatrix}$，$X(:,:,2) = \begin{bmatrix} 7 & -1 \end{bmatrix}$，则 $Y(:,:,1) = \begin{bmatrix} 1 & 2 \end{bmatrix}$，$Y(:,:,2) = \begin{bmatrix} 3 & 4 \end{bmatrix}$

图 3.6 所示为一维离散傅里叶反变换结果。图 3.6(a) 为频谱图像，图 3.6(b) 为频谱图像经过一维傅里叶反变换（列反变换）后的灰度图像。

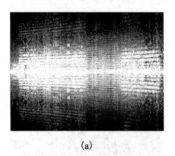 

(a)   (b)

图 3.6   一维离散傅里叶反变换

(2) Y = ifft(X, n)，通过快速算法进行一维离散傅里叶反变换，其中 n 为指定的变换点数。

(3) Y = ifft(X, [], dim) 和 Y = ifft(X, n, dim)，沿第 dim 维方向进行离散傅里叶反变换。

图 3.7 所示为一维离散傅里叶反变换结果。图 3.7(a) 为频谱图像，图 3.7(b) 为频谱图像经过一维傅里叶反变换（沿第二维方向反变换）后的灰度图像。

2. 二维离散傅里叶变换

MATLAB 利用 fft2 函数通过快速算法实现二维离散傅里叶变换，其主要用法如下：

(1) Y = fft2(X)，通过快速算法进行二维离散傅里叶变换，其中 X 和 Y 具有相同尺寸。例如：

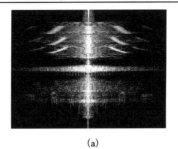

图 3.7 一维离散傅里叶反变换

$$X = \begin{bmatrix} 1 & 2 \\ 3 & 4 \end{bmatrix}, \text{则 } Y = \begin{bmatrix} 10 & -2 \\ -4 & 0 \end{bmatrix}$$

(2) Y = fft2(X, m, n)，通过快速算法进行二维离散傅里叶变换，其中 m 和 n 表示变换前把 X 截断或填充到尺寸为 m×n 的矩阵。变换后的矩阵尺寸为 m×n。

图 3.8 所示为二维离散傅里叶变换结果。图 3.8(a) 为原始灰度图像，图 3.8(b) 为原始灰度图像经过二维傅里叶变换后的频谱图像。

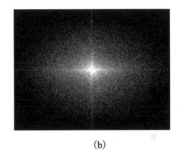

图 3.8 二维离散傅里叶变换

MATLAB 利用 ifft2 函数通过快速算法实现二维离散傅里叶反变换，其主要用法如下：

(1) Y = ifft2(X)，通过快速算法进行二维离散傅里叶反变换，其中 X 和 Y 具有相同尺寸。例如：

$$X = \begin{bmatrix} 10 & -2 \\ -4 & 0 \end{bmatrix}, \text{则 } Y = \begin{bmatrix} 1 & 2 \\ 3 & 4 \end{bmatrix}$$

(2) Y = ifft2(X, m, n)，通过快速算法进行二维离散傅里叶反变换，其中 m 和 n 表示返回矩阵的尺寸为 m×n。

图 3.9 所示为二维离散傅里叶反变换结果。图 3.9(a) 为频谱图像，图 3.9(b) 为频谱图像经过二维傅里叶反变换后的灰度图像。

3. 多维离散傅里叶变换

MATLAB 利用 fftn 函数通过快速算法实现多维离散傅里叶变换，其主要用法如下：

(1) Y = fftn(X)，通过快速算法进行多维离散傅里叶变换，其中 X 和 Y 具有相同尺寸。例如：

(a)   (b)

图 3.9   二维离散傅里叶反变换

$$X(:,:,1) = \begin{bmatrix} 1 & 2 \\ 3 & 4 \end{bmatrix}, \quad X(:,:,2) = \begin{bmatrix} 5 & 6 \\ 7 & 8 \end{bmatrix}, \quad 则\ Y(:,:,1) = \begin{bmatrix} 36 & -4 \\ -8 & 0 \end{bmatrix}, \quad Y(:,:,2) = \begin{bmatrix} -16 & 0 \\ 0 & 0 \end{bmatrix}$$

(2) Y = fftn(X, siz)，通过快速算法进行多维离散傅里叶变换，其中 siz 表示变换前把 X 截断或填充到尺寸为 siz 的多维数组。变换后的多维数组尺寸为 siz。

MATLAB 利用 ifftn 函数通过快速算法实现多维离散傅里叶反变换，其主要用法如下：

(1) Y = ifftn(X)，通过快速算法进行多维离散傅里叶反变换，其中 X 和 Y 具有相同尺寸。例如：

$$X(:,:,1) = \begin{bmatrix} 36 & -4 \\ -8 & 0 \end{bmatrix}, \quad X(:,:,2) = \begin{bmatrix} -16 & 0 \\ 0 & 0 \end{bmatrix}, \quad 则\ Y(:,:,1) = \begin{bmatrix} 1 & 2 \\ 3 & 4 \end{bmatrix}, \quad Y(:,:,2) = \begin{bmatrix} 5 & 6 \\ 7 & 8 \end{bmatrix}$$

(2) Y = ifftn(X, siz)，通过快速算法进行多维离散傅里叶反变换，其中 siz 表示变换前把 X 截断或填充到尺寸为 siz 的多维数组。变换后的多维数组尺寸为 siz。

4. 频谱移动

MATLAB 利用 fftshift 函数把零频分量移到频谱中心，其主要用法如下：

(1) Y = fftshift(X)，重新排列 fft、fft2 和 fftn 等函数的输出频谱，把零频分量移到频谱中心。如果 X 为矢量，fftshift(X) 把 X 的左半与右半交换；如果 X 为矩阵，fftshift(X) 把 X 的第一象限与第三象限交换，第二象限与第四象限交换；如果 X 为高维数组，fftshift(X) 在每一维方向把 X 的两个半空间交换。例如：

$$X = \begin{bmatrix} 1 & 2 & 3 & 4 & 5 \end{bmatrix}, \quad 则\ Y = \begin{bmatrix} 4 & 5 & 1 & 2 & 3 \end{bmatrix}$$

$$X = \begin{bmatrix} 1 & 2 & 3 & 4 & 5 \\ 6 & 7 & 8 & 9 & 10 \end{bmatrix}, \quad 则\ Y = \begin{bmatrix} 9 & 10 & 6 & 7 & 8 \\ 4 & 5 & 1 & 2 & 3 \end{bmatrix}$$

$$X(:,:,1) = \begin{bmatrix} 1 & 2 & 3 \\ 4 & 5 & 6 \end{bmatrix}, \quad X(:,:,2) = \begin{bmatrix} 7 & 8 & 9 \\ 10 & 11 & 12 \end{bmatrix},$$

$$则\ Y(:,:,1) = \begin{bmatrix} 12 & 10 & 11 \\ 9 & 7 & 8 \end{bmatrix}, \quad Y(:,:,2) = \begin{bmatrix} 6 & 4 & 5 \\ 3 & 1 & 2 \end{bmatrix}$$

(2) Y = fftshift(X, dim)，在第 dim 维方向进行 fftshift 操作。

MATLAB 利用 ifftshift 函数把零频分量从频谱中心移开，其主要用法如下：

(1) Y = ifftshift(X)，把矢量 X 的左半与右半交换；对于矩阵 X，ifftshift(X) 把 X 的第一象限与第三象限交换，第二象限与第四象限交换；如果 X 是高维数组，ifftshift(X) 在每一维方向把 X 的两个半空间交换。例如：

$$X = \begin{bmatrix} 1 & 2 & 3 & 4 & 5 \end{bmatrix}，则 Y = \begin{bmatrix} 3 & 4 & 5 & 1 & 2 \end{bmatrix}$$

$$X = \begin{bmatrix} 1 & 2 & 3 & 4 & 5 \\ 6 & 7 & 8 & 9 & 10 \end{bmatrix}，则 Y = \begin{bmatrix} 8 & 9 & 10 & 6 & 7 \\ 3 & 4 & 5 & 1 & 2 \end{bmatrix}$$

$$X(:,:,1) = \begin{bmatrix} 1 & 2 & 3 \\ 4 & 5 & 6 \end{bmatrix}，X(:,:,2) = \begin{bmatrix} 7 & 8 & 9 \\ 10 & 11 & 12 \end{bmatrix},$$

$$则 Y(:,:,1) = \begin{bmatrix} 11 & 12 & 10 \\ 8 & 9 & 7 \end{bmatrix}，Y(:,:,2) = \begin{bmatrix} 5 & 6 & 4 \\ 2 & 3 & 1 \end{bmatrix}$$

(2) Y = ifftshift(X, dim)，在第 dim 维方向进行 ifftshift 操作。

注意，对于行列均为偶数矩阵 X，则 ifftshift(fftshift(X)) = fftshift(ifftshift(X)) = X；否则 ifftshift(fftshift(X)) = X，但 fftshift(ifftshift(X)) ≠ X。

3.1.6 离散傅里叶变换应用

图 3.10 所示为离散傅里叶变换应用。图 3.10(a) 和图 3.10(b) 为对应于物体变形前后的两幅单曝光数字散斑图；图 3.10(c) 为散斑图经过离散傅里叶变换后得到的傅里叶频谱分布；图 3.10(d) 为所设计的理想带通滤波器；图 3.10(e) 为通过带通滤波后的傅里叶频谱分布；图 3.10(f) 为经过离散傅里叶反变换后得到的离面位移导数的等值条纹图。

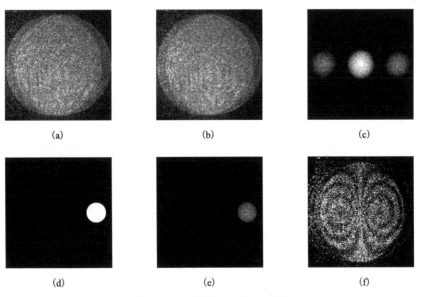

图 3.10 离散傅里叶变换应用

3.2 余弦变换

尽管傅里叶变换在信号处理和图像处理中获得了广泛应用，但傅里叶变换要涉及复数运算，因此运算量较大，复数运算量相当于实数的两倍。为了克服傅里叶变换的上述问题，人们提出了余弦变换(cosine transform)。余弦变换是以一组不同频率和不同幅值的余弦函数之和来近似表征一幅图像，实际上它是傅里叶变换的实数部分，因此余弦变换的运算量较小。

3.2.1 离散余弦变换

1. 一维离散余弦变换

设 $f(x)$ 为一维实数离散序列，其中 $x=0,1,\cdots,M-1$，则一维离散余弦变换定义为

$$F(u) = C(u) \sum_{x=0}^{M-1} f(x) \cos\left[\frac{\pi(2x+1)u}{2M}\right] \quad (u=0,1,\cdots,M-1) \tag{3.34}$$

式中，u 为离散频率变量，$C(u) = \begin{cases} 1/\sqrt{M} & (u=0) \\ \sqrt{2/M} & (u=1,2,\cdots,M-1) \end{cases}$。

一维离散余弦反变换定义为

$$f(x) = C(x) \sum_{u=0}^{M-1} F(u) \cos\left[\frac{\pi(2x+1)u}{2M}\right] \quad (x=0,1,\cdots,M-1) \tag{3.35}$$

式中

$$C(x) = \begin{cases} 1/\sqrt{M} & (x=0) \\ \sqrt{2/M} & (x=1,2,\cdots,M-1) \end{cases}$$

显然，对于离散余弦变换，其变换和反变换具有相同的变换核。

2. 二维离散余弦变换

设 $f(x,y)$ 为二维实数离散序列，其中 $x=0,1,\cdots,M-1$；$y=0,1,\cdots,N-1$，则二维离散余弦变换定义为

$$F(u,v) = C(u)C(v) \sum_{x=0}^{M-1} \sum_{y=0}^{N-1} f(x,y) \cos\left[\frac{\pi(2x+1)u}{2M}\right] \cos\left[\frac{\pi(2y+1)v}{2N}\right] \tag{3.36}$$
$$(u=0,1,\cdots,M-1; v=0,1,\cdots,N-1)$$

式中

$$C(u) = \begin{cases} 1/\sqrt{M} & (u=0) \\ \sqrt{2/M} & (u=1,2,\cdots,M-1) \end{cases}$$

$$C(v) = \begin{cases} 1/\sqrt{N} & (v=0) \\ \sqrt{2/N} & (v=1,2,\cdots,N-1) \end{cases}$$

二维离散余弦反变换定义为

$$F(x,y) = C(x)C(y) \sum_{u=0}^{M-1} \sum_{v=0}^{N-1} f(u,v) \cos\left[\frac{\pi(2x+1)u}{2M}\right] \cos\left[\frac{\pi(2y+1)v}{2N}\right] \quad (3.37)$$

$$(x = 0,1,\cdots,M-1; y = 0,1,\cdots,N-1)$$

式中

$$C(x) = \begin{cases} 1/\sqrt{M} & (x=0) \\ \sqrt{2/M} & (x=1,2,\cdots,M-1) \end{cases}$$

$$C(y) = \begin{cases} 1/\sqrt{N} & (y=0) \\ \sqrt{2/N} & (y=1,2,\cdots,N-1) \end{cases}$$

3.2.2 离散余弦变换算法

1. 一维离散余弦变换

MATLAB 利用 dct 函数实现一维离散余弦变换，其主要用法如下：

(1) Y = dct(X)，进行一维离散余弦变换，其中 X 和 Y 具有相同尺寸。如果 X 是矩阵，那么 dct 函数对矩阵每列进行余弦变换。例如：

$$X = \begin{bmatrix} 1 & 2 \end{bmatrix},\ \text{则}\ Y = \begin{bmatrix} 2.1213 & -0.7071 \end{bmatrix}$$

$$X = \begin{bmatrix} 1 & 2 \\ 3 & 4 \end{bmatrix},\ \text{则}\ Y = \begin{bmatrix} 2.8284 & 4.2426 \\ -1.4142 & -1.4142 \end{bmatrix}$$

图 3.11 所示为一维离散余弦变换结果。图 3.11(a) 为原始灰度图像，图 3.11(b) 为原始灰度图像经过一维余弦变换(列变换)后的频谱图像。

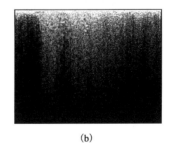

(a)　　　　　　　　(b)

图 3.11　一维离散余弦变换

(2) Y = dct(X, n)，进行一维离散余弦变换，其中 n 为指定的变换点数。

MATLAB 利用 idct 函数实现一维离散余弦反变换，其主要用法如下：

(1) Y = idct(X)，进行一维离散余弦反变换，其中 X 和 Y 具有相同尺寸。如果 X 是矩阵，那么 idct 函数对矩阵每列进行余弦反变换。例如：

$$X = \begin{bmatrix} 2.1213 & -0.7071 \end{bmatrix},\ \text{则}\ Y = \begin{bmatrix} 1.0000 & 2.0000 \end{bmatrix}$$

$$X = \begin{bmatrix} 2.8284 & 4.2426 \\ -1.4142 & -1.4142 \end{bmatrix},\ \text{则}\ Y = \begin{bmatrix} 1.0000 & 2.0000 \\ 3.0000 & 4.0000 \end{bmatrix}$$

图 3.12 所示为一维离散余弦反变换结果。图 3.12(a) 为频谱图像，图 3.12(b) 为频谱图像经过一维余弦反变换(列反变换)后的灰度图像。

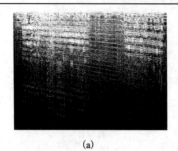 

(a)   (b)

图 3.12   一维离散余弦反变换

(2) Y = idct(X, n)，进行一维离散余弦反变换，其中 n 为指定的变换点数。

2. 二维离散余弦变换

MATLAB 利用 dct2 函数实现二维离散余弦变换，其主要用法如下：

(1) Y = dct2(X)，进行二维离散余弦变换，其中 X 和 Y 具有相同尺寸。例如：

$$X = \begin{bmatrix} 1 & 2 \\ 3 & 4 \end{bmatrix}, \text{则} Y = \begin{bmatrix} 5.0000 & -1.0000 \\ -2.0000 & 0 \end{bmatrix}$$

图 3.13 所示为二维离散余弦变换结果。图 3.13(a)为原始灰度图像，图 3.13(b)为原始灰度图像经过二维余弦变换后的频谱图像。

 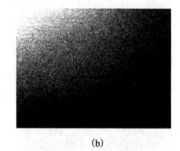

(a)   (b)

图 3.13   二维离散余弦变换

(2) Y = dct2(X, m, n) 或 Y = dct2(X, [m n])，进行二维离散余弦变换，其中 m 和 n 表示变换前把 X 截断或填充 0 到尺寸为 m×n 的矩阵。变换后的矩阵尺寸为 m×n。

MATLAB 利用 idct2 函数实现二维离散余弦反变换，其主要用法如下：

(1) Y = idct2(X)，进行二维离散余弦变换，其中 X 和 Y 具有相同尺寸。例如：

$$X = \begin{bmatrix} 5.0000 & -1.0000 \\ -2.0000 & 0 \end{bmatrix}, \text{则} Y = \begin{bmatrix} 1.0000 & 2.0000 \\ 3.0000 & 4.0000 \end{bmatrix}$$

图 3.14 所示为二维离散余弦反变换结果。图 3.14(a)为频谱图像，图 3.14(b)为频谱图像经过二维余弦反变换后的灰度图像。

(2) Y = idct2(X, m, n) 或 Y = idct2(X, [m n])，进行二维离散余弦反变换，其中 m 和 n 表示变换前把 X 截断或填充 0 到尺寸为 m×n 的矩阵。变换后的矩阵尺寸为 m×n。

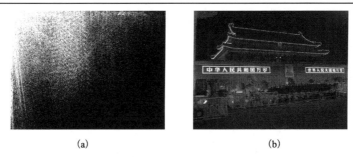

图 3.14 二维离散余弦反变换

3.2.3 离散余弦变换应用

图 3.15 所示为离散余弦变换应用。图 3.15(a)和图 3.15(b)为对应于物体变形前后的两幅单曝光数字散斑图;图 3.15(c)为数字散斑图经过离散余弦变换后得到的余弦频谱分布;图 3.15(d)为带通滤波器;图 3.15(e)为经过滤波后的频谱分布;图 3.15(f)为经过离散余弦反变换后得到的离面位移导数的等值条纹图。

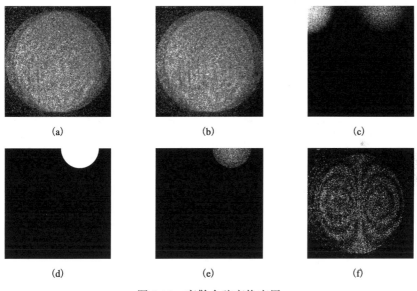

图 3.15 离散余弦变换应用

3.3 小波变换

傅里叶变换自提出以来一直是信号分析和图像处理的重要工具。然而,傅里叶变换无法处理非平稳信息。因此,寻找新的变换技术,使之能够处理非平稳信息就成为新的研究热点。小波变换(wavelet transform)正是在这样的需求背景下发展起来的变换技术。

与傅里叶变换相比,小波变换是时(空)域和频域的局域变换,能更有效地提取和分析局部信息。小波变换能够将信号和图像按小波分解,根据需要确定分解层次,可以有效控制计算量。另外，小波变换具有缩放(scaling)和平移(shifting)功能，可以产生不同尺度信号和图像。小波变换因具有上述优点而在信号和图像处理领域具有广泛应用。

3.3.1 连续小波变换

1. 一维连续小波变换

小波(wavelet)是通过对基本小波进行缩放和平移而得到的。基本小波是指均值为零的有限振荡波形，满足条件

$$\int_{-\infty}^{\infty}\psi(x)\mathrm{d}x=0, \quad C_\psi = \int_{-\infty}^{\infty}\frac{|\hat{\psi}(\omega)|^2}{|\omega|}\mathrm{d}\omega < \infty \tag{3.38}$$

式中，$\hat{\psi}(\omega)=\int_{-\infty}^{\infty}\psi(x)\exp(-\mathrm{i}\omega x)\mathrm{d}x$，$C_\psi$是与$\psi$有关的常数。一维连续小波函数可表示为

$$\psi_{a,b}(x)=\frac{1}{\sqrt{a}}\psi\left(\frac{x-b}{a}\right) \quad (a,b\in R, a>0) \tag{3.39}$$

一维连续小波变换(continuous wavelet transform)定义为

$$W(a,b)=\int_{-\infty}^{\infty}f(x)\psi_{a,b}^*(x)\mathrm{d}x \quad (a,b\in R, a>0) \tag{3.40}$$

式中，*表示复共轭。其反变换(inverse continuous wavelet transform)定义为

$$f(x)=\frac{1}{C_\psi}\int_0^\infty\int_{-\infty}^{\infty}W(a,b)\psi_{a,b}(x)\frac{\mathrm{d}a\mathrm{d}b}{a^2} \tag{3.41}$$

式中，$C_\psi=\int_{-\infty}^{\infty}\frac{|\hat{\psi}(\omega)|^2}{|\omega|}\mathrm{d}\omega$。

2. 二维连续小波变换

二维连续小波函数定义为

$$\psi_{a,b,c}(x,y)=\frac{1}{a}\psi\left(\frac{x-b}{a},\frac{y-c}{a}\right) \quad (a,b,c\in R, a>0) \tag{3.42}$$

二维连续小波变换定义为

$$W(a,b,c)=\int_{-\infty}^{\infty}f(x,y)\psi_{a,b,c}^*(x,y)\mathrm{d}x\mathrm{d}y \quad (a,b,c\in R, a>0) \tag{3.43}$$

其反变换定义为

$$f(x,y)=\frac{1}{C_\psi}\int_0^\infty\int_{-\infty}^{\infty}\int_{-\infty}^{\infty}W(a,b,c)\psi_{a,b,c}(x,y)\frac{\mathrm{d}a\mathrm{d}b\mathrm{d}c}{a^3} \tag{3.44}$$

式中，$C_\psi=\int_{-\infty}^{\infty}\int_{-\infty}^{\infty}\frac{|\hat{\psi}(\omega_1,\omega_2)|^2}{|\omega_1^2+\omega_2^2|}\mathrm{d}\omega_1\mathrm{d}\omega_2$。

3.3.2 离散小波变换

1. 一维离散小波变换

设尺度因子 $a=2^j$ 和平移因子 $b=ka=k2^j$，则一维离散小波函数可表示为

$$\psi_{j,k}(x)=\frac{1}{\sqrt{2^j}}\psi\left(\frac{1}{2^j}x-k\right) \quad (j,k\in Z) \tag{3.45}$$

式中，$\psi_{0,0}(x)=\psi(x)$。则一维离散小波变换(discrete wavelet transform)定义为

$$W(j,k)=\int_{-\infty}^{\infty}f(x)\psi_{j,k}^*(x)\mathrm{d}x \quad (j,k\in Z) \tag{3.46}$$

其反变换(inverse discrete wavelet transform)定义为

$$f(x)=\sum_{-\infty}^{\infty}\sum_{-\infty}^{\infty}W(j,k)\psi_{j,k}(x) \quad (j,k\in Z) \tag{3.47}$$

2. 二维离散小波变换

将一维离散小波变换推广到二维，即可得到二维离散小波变换。设 $\varphi_{j,k}(x)$ 和 $\varphi_{j,l}(y)$ 是一维离散尺度函数，$\psi_{j,k}(x)$ 和 $\psi_{j,l}(y)$ 是相应的一维离散小波函数，则二维离散小波函数为

$$\begin{aligned}\varphi_A(x,y)&=\varphi_{j,k}(x)\varphi_{j,l}(y)\\ \psi_H(x,y)&=\varphi_{j,k}(x)\psi_{j,l}(y)\\ \psi_V(x,y)&=\psi_{j,k}(x)\varphi_{j,l}(y)\\ \psi_D(x,y)&=\psi_{j,k}(x)\psi_{j,l}(y)\end{aligned} \quad (j,k,l\in Z) \tag{3.48}$$

式中，$\varphi_A(x,y)$ 为二维离散尺度函数，$\psi_H(x,y),\psi_V(x,y)$ 和 $\psi_D(x,y)$ 为分别与水平、垂直和对角细节对应的二维离散小波函数。

二维离散小波变换可以转化为两次一维离散小波变换，首先对图像矩阵的各行进行离散小波变换，然后再对各列进行离散小波变换。经过两次一维离散小波变换，即可实现二维离散小波变换，如图 3.16 所示。二维离散小波变换将二维数字图像在不同尺度上进行分解(decomposition)，分解结果包括近似分量 cA、水平细节分量 cH、垂直细节分量 cV 和对角细节分量 cD。

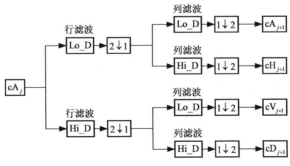

图 3.16 二维小波分解示意图

图中，$\boxed{2\downarrow1}$表示对列降样（downsample columns），保留偶数列；$\boxed{1\downarrow2}$表示对行降样（downsample rows），保留偶数行。

二维离散小波反变换是利用二维小波变换结果在不同尺度上进行二维数字图像重构（reconstruction）。二维离散小波反变换也可以转化为两次一维离散小波反变换，首先对图像矩阵的各列进行离散小波反变换，然后再对各行进行离散小波反变换。经过两次一维离散小波反变换，即可实现二维离散小波反变换，如图3.17所示。

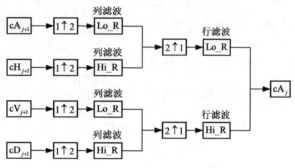

图 3.17　二维小波重构示意图

图中，$\boxed{1\uparrow2}$表示对行升样（upsample rows），奇数行插入 0；$\boxed{2\uparrow1}$表示对列升样（upsample columns），奇数列插入 0。

对二维图像进行单层（single-level）二维离散小波变换，可产生 1 个近似（approximation）子图 cA1 和 3 个细节（detail）子图（即水平细节 cH1、垂直细节 cV1 和对角细节 cD1）。对图像进行单层二维离散小波分解后的子图分布如图 3.18 所示。cA1 子图反映原图的低频信息，cH1 子图反映原图水平细节的高频信息，cV1 子图反映原图垂直细节的高频信息，cD1 子图反映原图对角细节的高频信息。

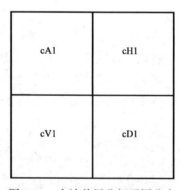

图 3.18　小波单层分解子图分布

图像的下层小波变换是在前层产生的低频子图的基础上进行的，依次重复即可完成图像的多层（multi-level）二维离散小波分解。由于对图像每进行一层小波变换，就相当于在水平和垂直方向分别进行隔点采样，因此变换后的图像就分解为 4 个大小为前层图像 1/4 尺寸的频带子图。对图像进行三层二维离散小波分解后的子图分布如图 3.19 所示。

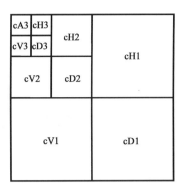

图 3.19 小波三层分解子图分布

3.3.3 离散小波变换算法

1. 单层一维离散小波变换

MATLAB 利用 dwt 函数进行单层一维离散小波分解，其主要用法如下：

(1) [cA, cD] = dwt(X, 'wname')，通过输入矢量 X 的一维小波分解计算近似系数矢量 cA 和细节系数矢量 cD，其中小波名 wname 可选如下：

(a) Daubechies: db1/haar, db2, ⋯, db45

(b) Coiflets: coif1, coif2, ⋯, coif5

(c) Symlets: sym2, sym3, ⋯, sym45

(d) Discrete Meyer: dmey

(e) Biorthogonal: bior1.1, bior1.3, bior1.5, bior2.2, bior2.4, bior2.6, bior2.8, ⋯

(f) Reverse Biorthogonal: rbio1.1, rbio1.3, rbio1.5, rbio2.2, rbio2.4, rbio2.6, rbio2.8, ⋯

(2) [cA, cD] = dwt(X, Lo_D, Hi_D)，采用指定的滤波器进行一维小波分解，其中 Lo_D 和 Hi_D 分别为进行小波分解的低通和高通滤波器(Lo_D 和 Hi_D 必须具有相同长度)。

MATLAB 利用 idwt 函数进行单层一维离散小波重构，其主要用法如下：

(1) X = idwt(cA, cD, 'wname')，基于近似系数矢量 cA 和细节系数矢量 cD，采用名为 wname 的小波计算单层重构近似系数矢量 X。

(2) X = idwt(cA, cD, Lo_R, Hi_R)，采用指定的滤波器进行一维小波重构，其中 Lo_D 和 Hi_D 分别为进行小波重构的低通和高通滤波器(Lo_D 和 Hi_D 必须具有相同长度)。

(3) X = idwt(cA, [], ⋯)，基于近似系数矢量 cA 计算单层重构近似系数矢量 X。

(4) X = idwt2([], cD, ⋯)，基于细节系数矢量 cD 计算单层重构细节系数矢量 X。

2. 单层二维离散小波变换

MATLAB 利用 dwt2 函数进行单层二维离散小波分解，其主要用法如下：

(1) [cA, cH, cV, cD] = dwt2(X, 'wname')，通过输入矩阵 X 的二维小波分解计算近似系数矩阵 cA(近似分量)和细节系数矩阵 cH(水平分量)、cV(垂直分量)和 cD(对角分量)，其中 wname 是指小波名。

图 3.20 所示为单层二维离散小波变换结果。图 3.20(a)为原始灰度图像，图 3.20(b)

为原始灰度图像经过单层二维小波变换后的近似系数(cA)和细节系数(cH、cV 和 cD)图像。

(a) （b)

图 3.20　二维离散小波变换

(2) [cA, cH, cV, cD] = dwt2(X, Lo_D, Hi_D)，采用指定的小波分解滤波器进行二维小波分解，其中 Lo_D 和 Hi_D 分别为进行小波分解的低通和高通滤波器(Lo_D 和 Hi_D 必须具有相同长度)。

MATLAB 利用 idwt2 函数进行单层二维离散小波重构，其主要用法如下：

(1) X = idwt2(cA, cH, cV, cD, 'wname')，基于近似系数矩阵 cA(近似分量)和细节系数矩阵 cH(水平分量)、cV(垂直分量)和 cD(对角分量)，采用名为 wname 的小波计算单层重构近似系数矩阵 X。

图 3.21 所示为单层二维离散小波反变换结果。图 3.21(a)为近似系数(cA)和细节系数(cH、cV 和 cD)图像，图 3.21(b)为近似系数(cA)和细节系数(cH、cV 和 cD)图像经过单层二维小波反变换后的灰度图像。

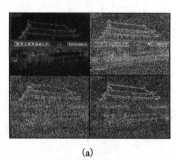 

(a) （b)

图 3.21　二维离散小波反变换

(2) X = idwt2(cA, cH, cV, cD, Lo_R, Hi_R)，采用指定的滤波器进行二维小波重构，其中 Lo_D 和 Hi_D 分别为进行小波重构的低通和高通滤波器(Lo_D 和 Hi_D 必须具有相同长度)。

(3) X = idwt2(cA, [], [], [], …)，基于近似系数矩阵 cA 计算单层重构近似系数矩阵 X。

图 3.22 所示为单层二维离散小波反变换结果。图 3.22(a)为近似系数(cA)图像，图 3.22(b)为近似系数(cA)图像经过单层二维小波反变换后的灰度图像。

(4) X = idwt2([], cH, [], [], …)、X = idwt2([], [], cV, [], …)和 X = idwt2([], [], [], cD, …)，分别为基于水平、垂直和对角细节系数矩阵 cH、cV 和 cD 计算单层重构细节系数矩阵 X。

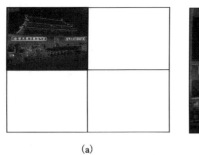 

图 3.22　二维离散小波反变换

图 3.23 所示为单层二维离散小波反变换结果。图 3.23(a)为细节系数(cH、cV 和 cD)图像，图 3.23(b)为细节系数(cH、cV 和 cD)图像经过单层二维小波反变换后的灰度图像。

图 3.23　二维离散小波反变换

3. 多层一维离散小波变换

MATLAB 利用 wavedec 函数进行多层一维离散小波分解，其主要用法如下：

(1) [C, L] = wavedec(X, N, 'wname')，利用名为 wname 的小波进行矢量 X 的 N 级小波分解，其中 C 为分解矢量、L 为簿记矢量、N 为正整数。

(2) [C, L] = wavedec(X, N, Lo_D, Hi_D)，采用指定的小波分解滤波器进行一维小波分解，其中 Lo_D 和 Hi_D 分别为进行小波分解的低通和高通滤波器(Lo_D 和 Hi_D 必须具有相同长度)。

MATLAB 利用 waverec 函数进行多层一维离散小波重构，其主要用法如下：

(1) X = waverec(C, L, 'wname')，基于小波分解结构[C, L]进行矢量 X 的多层小波重构，其中 wname 是小波名。

(2) X = waverec(C, L, Lo_R, Hi_R)，基于小波分解结构[C, L]进行矢量 X 的多层小波重构，其中 Lo_R 和 Hi_R 分别为进行小波重构的低通和高通滤波器。

4. 多层二维离散小波变换

MATLAB 利用 wavedec2 函数进行多层二维离散小波分解，其主要用法如下：

(1) [C, S] = wavedec2(X, N, 'wname')，利用名为 wname 的小波进行矩阵 X 的 N 级小波分解。分解矢量 C 排列为 C = [A(N) H(N) V(N) D(N) H(N−1) V(N−1) D(N−1) …

H(1) V(1) D(1)]，其中 A 为近似系数、H 为水平细节系数、V 为垂直细节系数和 D 为对角细节系数。相应的簿记矩阵 S：S(1,:)为 N 级分解的近似系数矩阵尺寸；S(i,:)为在 N 级分解中第 i 级分解的细节系数的尺寸。N 必须是正整数。

(2)[C, S] = wavedec2(X, N, Lo_D, Hi_D)，采用指定的小波分解滤波器进行二维小波分解，其中 Lo_D 和 Hi_D 分别为进行小波分解的低通和高通滤波器(Lo_D 和 Hi_D 必须具有相同长度)。

MATLAB 利用 waverec2 函数进行多层二维离散小波重构，其主要用法如下：

(1) X = waverec2(C, S, 'wname')，基于小波分解结构[C, S]进行矩阵 X 的多层小波重构，其中 wname 是小波名。

(2) X = waverec2(C, S, Lo_R, Hi_R)，基于小波分解结构[C, S]进行矩阵 X 的多层小波重构，其中 Lo_R 和 Hi_R 分别为进行小波重构的低通和高通滤波器。

3.3.4 离散小波变换应用

图 3.24 所示为二维离散小波变换应用。图 3.24(a)和图 3.24(b)为对应于物体变形前后的两幅单曝光数字散斑图；图 3.24(c)为散斑图经过二维离散小波分解后的高频细节分量再经过二维离散小波重构后得到的离面位移导数的等值条纹图。

(a)　　　　　　　　　(b)　　　　　　　　　(c)

图 3.24　离散小波变换应用

第4章 图 像 降 噪

图像往往存在噪声(noise),因此通常需要进行低通滤波(low-pass filtering)以抑制或消除噪声。空域平滑滤波、频域低通滤波和小波低通滤波由于具有抑制或消除高频噪声的作用,因而已广泛应用于图像降噪(image denoising)。

4.1 空域平滑滤波

空域平滑滤波就是在空域对图像中的各个像素点的灰度值进行平滑处理,突出图像低频分量,抑制图像高频分量。空域平滑滤波包括均值滤波、中值滤波和自适应滤波。

4.1.1 均值滤波

1. 均值滤波原理

均值滤波是用具有奇数点的滑动窗口在图像上滑动,将窗口中心点对应的图像像素点的灰度值用窗口内的各个点的灰度值的平均值代替,窗口中灰度极高或极低的像素点对这种方法的影响很大,易造成边缘模糊。均值滤波实际上就是对输出像素邻域进行平均操作,再将平均值作为输出像素的灰度。均值滤波先要构造一个滤波模板,然后利用模板对图像进行滤波处理。

设 S 为包含像素 (x_0, y_0) 在内的邻域集合,(x, y) 为集合 S 中的像素,$f(x, y)$ 为像素 (x, y) 处的灰度值,则均值滤波后在像素 (x_0, y_0) 处的灰度值可表示为

$$g(x_0, y_0) = \frac{\sum_{(x,y) \in S} h(x,y) f(x,y)}{\sum_{(x,y) \in S} h(x,y)} \tag{4.1}$$

式中,$h(x, y)$ 为像素 (x, y) 处的权重。

2. 均值滤波算法

均值滤波可以通过相关或卷积实现。相关和卷积都是邻域操作,输出像素为其邻域输入像素的加权和。相关计算的权重矩阵称为相关核;卷积计算的权重矩阵称为卷积核。相关和卷积的主要差别在于权重矩阵不同,相关核在计算中并不进行旋转,而卷积核是由相关核旋转180°而得到。

MATLAB 利用 filter2 函数在空域对图像进行二维滤波,其主要用法如下:

(1)Y = filter2(h, X),通过二维相关在空域采用滤波器 h 对 X 进行二维滤波,其中 X 和 Y 具有相同尺寸。例如:

$$X = \begin{bmatrix} 11 & 12 & 13 & 14 & 15 & 16 \\ 21 & 22 & 23 & 24 & 25 & 26 \\ 31 & 32 & 73 & 84 & 35 & 36 \\ 41 & 42 & 43 & 44 & 45 & 46 \\ 51 & 52 & 53 & 54 & 55 & 56 \end{bmatrix}, \quad h = \frac{1}{9} \begin{bmatrix} 1 & 1 & 1 \\ 1 & 1 & 1 \\ 1 & 1 & 1 \end{bmatrix},$$

$$则\ Y = \begin{bmatrix} 7.3333 & 11.3333 & 12.0000 & 12.6667 & 13.3333 & 9.1111 \\ 14.3333 & 26.4444 & 33.0000 & 34.0000 & 30.5556 & 17.0000 \\ 21.0000 & 36.4444 & 43.0000 & 44.0000 & 40.5556 & 23.6667 \\ 27.6667 & 46.4444 & 53.0000 & 54.0000 & 50.5556 & 30.3333 \\ 20.6667 & 31.3333 & 32.0000 & 32.6667 & 33.3333 & 22.4444 \end{bmatrix}$$

(2) Y = filter2(h, X, shape)，通过二维相关在空域采用滤波器 h 对 X 进行二维滤波。如果 shape 指定为 full，则 Y 尺寸大于 X；如果 shape 指定为 same 或缺省，则 Y 尺寸等于 X；如果 shape 指定为 valid，则 Y 尺寸小于 X。

图 4.1 所示为均值滤波结果。图 4.1(a) 为原始灰度图像，图 4.1(b) 为原始灰度图像经过均值滤波后的灰度图像。

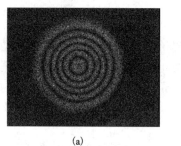
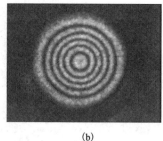

(a)　　　　　　　　　　　　(b)

图 4.1　均值滤波结果

3. 均值滤波应用

图 4.2 所示为采用三步相移而得到的干涉条纹。图 4.2(a)、图 4.2(b) 和图 4.2(c) 的相移量依次为 0, π/2 和 π。

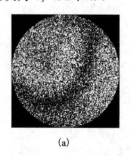
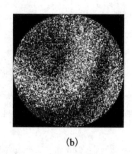
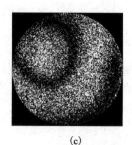

(a)　　　　　　　(b)　　　　　　　(c)

图 4.2　相移干涉条纹

图 4.3 所示为均值滤波应用。图 4.3(a)、图 4.3(b) 和图 4.3(c) 分别为图 4.2(a)、图 4.2(b)

和图 4.2(c)的均值滤波结果；图 4.3(d)和图 4.3(e)分别为相位在 $-\pi/2 \sim \pi/2$ 和 $0 \sim 2\pi$ 范围的包裹相位分布；图 4.3(f)为连续相位分布。

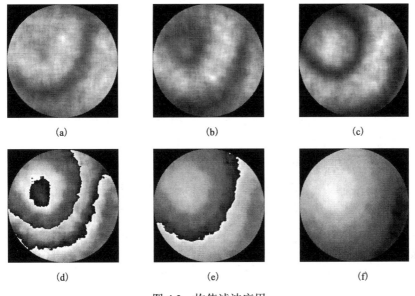

图 4.3　均值滤波应用

4.1.2　中值滤波

1. 中值滤波原理

中值滤波是用邻域内灰度的中值代替图像的当前点，能够在去除噪声的同时保留图像边缘细节，中值滤波不会产生明显的模糊边缘。中值滤波是非线性低通滤波方法，它可以有效保护图像边缘，同时可以去除噪声。与均值滤波不同，中值滤波是将邻域中的像素按灰度级排序，取中间值为输出像素。

设 S 为包含像素 (x_0,y_0) 在内的邻域集合，(x,y) 为集合 S 中的像素，$f(x,y)$ 为像素 (x,y) 处的灰度值，则中值滤波后在像素 (x_0,y_0) 处的灰度值可表示为

$$g(x_0,y_0)=\left[\underset{(x,y)\in S}{\mathrm{sort}}f(x,y)\right]_{\frac{m\times n+1}{2}} \tag{4.2}$$

式中，sort 表示排序，$m\times n$ 表示集合 S 中的像素个数，即滤波窗口大小。m 和 n 可以取不同值，这样可以得到不同滤波窗口。通常取奇数窗口，这样比较容易得到中值灰度值。中值滤波的效果与窗口的形状和大小有密切的关系。对二维图像，窗口的形状可以是矩形、圆形或十字形等，窗口的中心一般位于被处理点上。

2. 中值滤波算法

MATLAB 利用 medfilt2 函数在空域对图像进行二维数字滤波，其主要用法如下：

(1) B = medfilt2(A)，在空域采用 3×3 模板(即默认模板)对 X 进行二维滤波，其中 X 和 Y 具有相同尺寸。例如：

$$X = \begin{bmatrix} 11 & 12 & 13 & 14 & 15 & 16 \\ 21 & 22 & 23 & 24 & 25 & 26 \\ 31 & 32 & 73 & 84 & 35 & 36 \\ 41 & 42 & 43 & 44 & 45 & 46 \\ 51 & 52 & 53 & 54 & 55 & 56 \end{bmatrix}, 则 Y = \begin{bmatrix} 0 & 12 & 13 & 14 & 15 & 0 \\ 12 & 22 & 23 & 24 & 25 & 16 \\ 22 & 32 & 42 & 43 & 36 & 26 \\ 32 & 43 & 52 & 53 & 46 & 36 \\ 0 & 42 & 43 & 44 & 45 & 0 \end{bmatrix}$$

(2) Y = medfilt2(Y, [m n]), 在空域采用 m×n 模板对 X 进行二维滤波, 其中 X 和 Y 具有相同尺寸。

图 4.4 所示为中值滤波结果。图 4.4(a)为原始灰度图像, 图 4.4(b)为原始灰度图像经过中值滤波后的灰度图像。

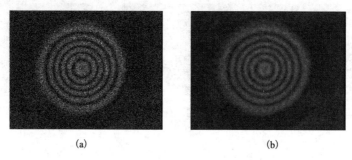

图 4.4 中值滤波结果

3. 中值滤波应用

图 4.5 所示为中值滤波应用。图 4.5(a)、图 4.5(b)和图 4.5(c)分别为图 4.2(a)、图 4.2(b)和图 4.2(c)的中值滤波结果; 图 4.5(d)和图 4.5(e)分别为相位在 −π/2 ～ π/2 和 0～2π 范围的包裹相位分布; 图 4.5(f)为连续相位分布。

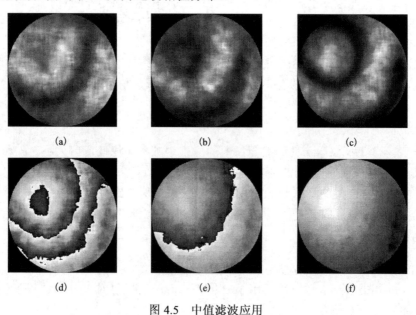

图 4.5 中值滤波应用

4.1.3 自适应滤波

1. 自适应滤波原理

自适应滤波是利用局部方差和图像噪声对输出像素邻域均值进行修正，将修正后的均值作为输出像素的灰度。自适应滤波先利用模板对图像进行均值滤波，然后再根据局部方差和图像噪声对均值进行修正。

设 S 为包含像素 (x_0, y_0) 在内的邻域集合，(x, y) 为集合 S 中的像素，$f(x, y)$ 为像素 (x, y) 处的灰度值，则自适应滤波后在像素 (x_0, y_0) 处的灰度值可表示为

$$g(x_0, y_0) = f(x, y) - \sigma^2 \frac{f(x, y) - \frac{1}{mn} \sum_{(x,y) \in S} f(x, y)}{\frac{1}{mn} \sum_{(x,y) \in S} f^2(x, y) - \left[\frac{1}{mn} \sum_{(x,y) \in S} f(x, y)\right]^2} \tag{4.3}$$

式中，$m \times n$ 表示集合 S 中的像素个数；σ^2 表示图像噪声。

2. 自适应滤波算法

MATLAB 利用 wiener2 函数对图像进行二维自适应滤波，其主要用法如下：

(1) [Y, noise] = wiener2(X, [m n])，采用 m×n 模板(默认时，[m n] = [3 3])对图像进行自适应滤波，其中 noise 是返回的滤波前图像的加性噪声。例如：

$$X = \begin{bmatrix} 11 & 12 & 13 & 14 & 15 & 16 \\ 21 & 22 & 23 & 24 & 25 & 26 \\ 31 & 32 & 73 & 84 & 35 & 36 \\ 41 & 42 & 43 & 44 & 45 & 46 \\ 51 & 52 & 53 & 54 & 55 & 56 \end{bmatrix}, \quad [m \ n] = [3 \ 3],$$

$$\text{则 } Y = \begin{bmatrix} 7.3333 & 11.3333 & 12.0000 & 12.6667 & 13.3333 & 9.1111 \\ 14.3333 & 26.4444 & 28.2198 & 29.3898 & 29.4382 & 17.0000 \\ 21.0000 & 36.4444 & 50.1243 & 52.0256 & 40.5556 & 23.6667 \\ 30.6744 & 46.4444 & 53.0000 & 54.0000 & 50.5556 & 35.7266 \\ 32.5847 & 38.5368 & 39.8655 & 41.1793 & 42.4792 & 38.6736 \end{bmatrix}, \quad \text{noise} = 330.9350$$

(2) J = wiener2(I, [m n], noise)，采用 m×n 模板对图像进行自适应滤波，其中 noise 是指定图像的加性噪声。

图 4.6 所示为自适应滤波结果。图 4.6(a) 为原始灰度图像，图 4.6(b) 为原始灰度图像经过自适应滤波后的灰度图像。

3. 自适应滤波应用

图 4.7 所示为自适应滤波应用。图 4.7(a)、图 4.7(b) 和图 4.7(c) 分别为图 4.2(a)、图 4.2(b) 和图 4.2(c) 的自适应滤波结果；图 4.7(d) 和图 4.7(e) 分别为相位在 $-\pi/2 \sim \pi/2$ 和 $0 \sim 2\pi$ 范围的包裹相位分布；图 4.7(f) 为连续相位分布。

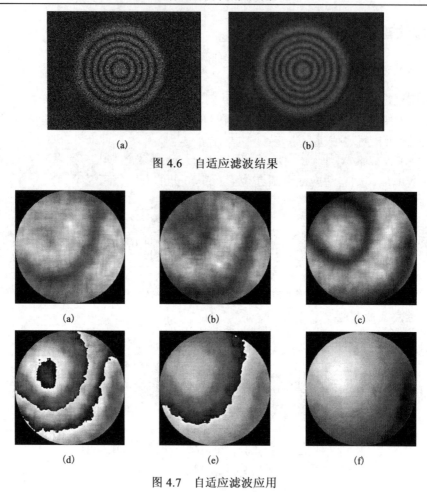

图 4.6 自适应滤波结果

图 4.7 自适应滤波应用

4.2 频域低通滤波

频域低通滤波就是在频域对图像进行低通滤波,然后再进行反变换,得到处理后的图像。对图像进行傅里叶变换和余弦变换就能得到它的频谱分布,图像缓变信息对应于低频分量,而图像突变信息(如图像细节、边缘等)则对应于高频分量。采用低通滤波可以突出图像缓变信息,抑制图像突变信息。

频域低通滤波主要包括理想低通滤波、巴特沃斯(Butterworth)低通滤波和指数低通滤波等。MATLAB 软件并未直接提供频域低通滤波的二维传递函数,但通过 MATLAB 语言可以容易编写这些滤波器的二维传递函数。

4.2.1 理想低通滤波

1. 理想低通滤波原理

理想低通滤波器定义为

$$H(u,v) = \begin{cases} 1 & (D(u,v) \leq D_0) \\ 0 & (D(u,v) > D_0) \end{cases} \tag{4.4}$$

式中，$H(u,v)$ 为传递函数；D_0 为截止频率；$D(u,v) = \sqrt{u^2 + v^2}$。

2. 理想低通滤波算法

利用 MATLAB 语言，频域二维理想低通滤波的传递函数构造如下：

(1) 离散傅里叶变换

```
H(1:M,1:N)=zeros;
for i=1:M
   for j=1:N
     if sqrt((i-(M+1)/2)^2+(j-(N+1)/2^)<=D
         H(i,j)=ones;
     end
   end
end
```

(2) 离散余弦变换

```
H(1:M,1:N)=zeros;
for i=1:M
   for j=1:N
     if sqrt(i^2+j^2)<=D
         H(i,j)=ones;
     end
   end
end
```

其中，M 和 N 分别为传递函数的行数和列数，D 为截止频率。

图 4.8 所示为理想低通滤波结果。图 4.8(a) 为原始灰度图像，图 4.8(b) 为原始灰度图像经过傅里叶变换理想低通滤波后的灰度图像，图 4.8(c) 为原始灰度图像经过余弦变换理想低通滤波后的灰度图像。

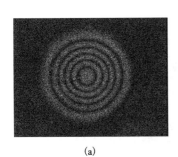 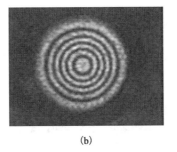 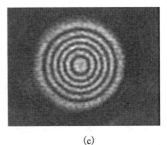

(a)　　　　　　　　　　(b)　　　　　　　　　　(c)

图 4.8　理想低通滤波结果

3. 理想低通滤波应用

图 4.9 和图 4.10 所示为理想低通滤波应用。图 4.9(a)、图 4.9(b) 和图 4.9(c) 分别为图 4.2(a)、图 4.2(b) 和图 4.2(c) 的傅里叶变换理想低通滤波结果，图 4.10(a)、图 4.10(b) 和图 4.10(c) 分别为图 4.2(a)、图 4.2(b) 和图 4.2(c) 的余弦变换理想低通滤波结果。图 4.9 和图 4.10 中的 (d)、(e) 分别为相位在 $-\pi/2 \sim \pi/2$ 和 $0 \sim 2\pi$ 范围的包裹相位分布，图 4.9(f) 和图 4.10(f) 为连续相位分布。

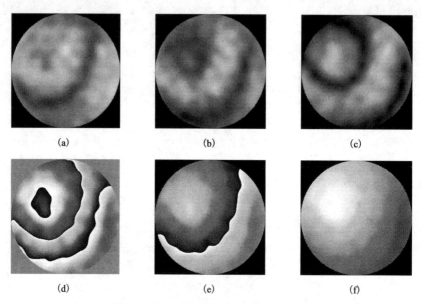

图 4.9 傅里叶变换理想低通滤波应用

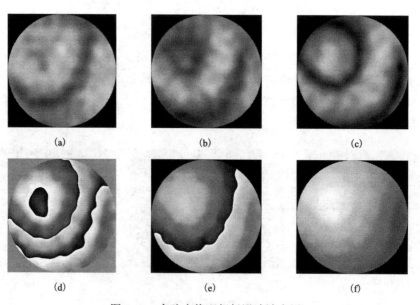

图 4.10 余弦变换理想低通滤波应用

4.2.2 巴特沃斯低通滤波

1. 巴特沃斯低通滤波原理

巴特沃斯低通滤波器定义为

$$H(u,v) = \frac{1}{1 + \left[\dfrac{D(u,v)}{D_0}\right]^{2n}} \tag{4.5}$$

式中，D_0 为截止频率；n 表示滤波器的阶数。当 $D(u,v) = D_0$ 时，$H(u,v) = 0.5$。

2. 巴特沃斯低通滤波算法

利用 MATLAB 语言，频域二维巴特沃斯低通滤波的传递函数构造如下：
(1) 离散傅里叶变换

```
for i=1:M
    for j=1:N
        H(i,j)=1/(1+(sqrt((i-)M+1)/2)^2+(j-(N+1)/2)^2)/D)^(2*n));
    end
end
```

(2) 离散余弦变换

```
for i=1:M
    for j=1:N
        H(i,j)=1/(1+(sqrt(i^2+j^2)/D)^(2*n));
    end
end
```

其中，M 和 N 分别为传递函数的行数和列数，D 为截止频率，n 为阶数。

图 4.11 所示为巴特沃斯低通滤波结果。图 4.11(a) 为原始灰度图像，图 4.11(b) 为原始灰度图像经过傅里叶变换巴特沃斯低通滤波后的灰度图像，图 4.11(c) 为原始灰度图像经过余弦变换巴特沃斯低通滤波后的灰度图像。

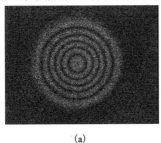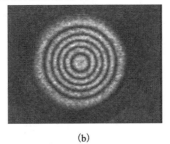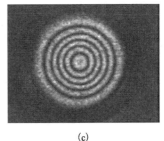

(a)　　　　　　　　　　(b)　　　　　　　　　　(c)

图 4.11　巴特沃斯低通滤波结果

3. 巴特沃斯低通滤波应用

图 4.12 和图 4.13 所示分别为巴特沃斯低通滤波应用，其中图 4.12 采用傅里叶变换，图 4.13 采用余弦变换。

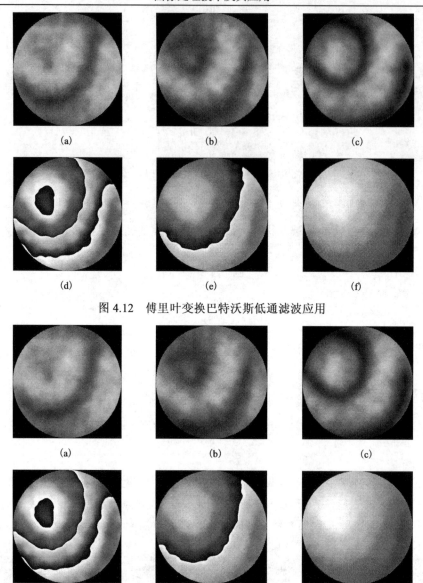

图 4.12 傅里叶变换巴特沃斯低通滤波应用

图 4.13 余弦变换巴特沃斯低通滤波应用

4.2.3 指数低通滤波

1. 指数低通滤波原理

指数低通滤波器定义为

$$H(u,v) = \exp\left\{-\left[\frac{D(u,v)}{D_0}\right]^n\right\} \tag{4.6}$$

式中，D_0 为截止频率；n 表示滤波器的衰减系数。当 $D(u,v) = D_0$ 时，$H(u,v) = 1/2.7$。

2. 指数低通滤波算法

利用 MATLAB 语言，频域二维指数低通滤波的传递函数构造如下：

(1) 离散傅里叶变换

```
for i=1:M
    for j=1:N
        H(i,j)=exp(-(sqrt((i-(M+1)/2)^2+(j-(N+1)/2)^2)/D)^n);
    end
end
```

(2) 离散余弦变换

```
for i=1:M
    for j=1:N
        H(i,j)=exp(-(sqrt(i^2+j^2)/D)^n);
    end
end
```

其中，M 和 N 分别为传递函数的行数和列数，D 为截止频率，n 为衰减系数。

图 4.14 所示为指数低通滤波结果。图 4.14(a)为原始灰度图像，图 4.14(b)为原始灰度图像经过傅里叶变换指数低通滤波后的灰度图像，图 4.14(c)为原始灰度图像经过余弦变换指数低通滤波后的灰度图像。

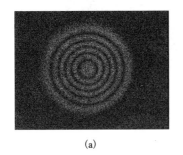 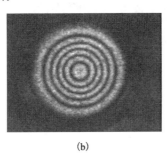 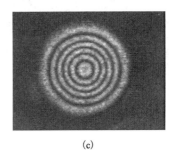

(a)　　　　　　　　　　　(b)　　　　　　　　　　　(c)

图 4.14　指数低通滤波结果

3. 指数低通滤波应用

图 4.15 和图 4.16 所示分别为指数低通滤波应用，其中图 4.15 采用傅里叶变换，图 4.16 采用余弦变换。

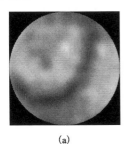 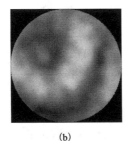 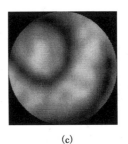

(a)　　　　　　　　　　　(b)　　　　　　　　　　　(c)

图 4.15　傅里叶变换指数低通滤波应用

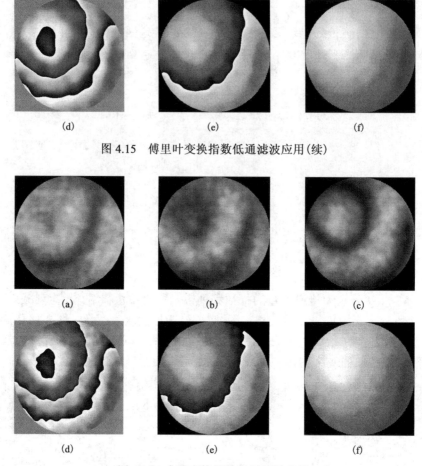

图 4.15 傅里叶变换指数低通滤波应用(续)

图 4.16 余弦变换指数低通滤波应用

4.3 同态低通滤波

4.3.1 同态低通滤波原理

一幅图像可以表示为

$$f(x,y) = i(x,y)r(x,y) \tag{4.7}$$

式中，$i(x,y)$ 和 $r(x,y)$ 分别为照明分量和反射分量。上式取自然对数，得

$$\ln f(x,y) = \ln i(x,y) + \ln r(x,y) \tag{4.8}$$

上式取傅里叶变换，得

$$FT[\ln f(x,y)] = FT[\ln i(x,y)] + FT[\ln r(x,y)] \tag{4.9}$$

式中，FT[…]表示傅里叶变换，$FT[\ln i(x,y)]$ 和 $FT[\ln r(x,y)]$ 分别为低频分量和高频分量。进行低通滤波，则上式可表示为

$$\text{FT}[\ln f'(x,y)] = \text{FT}[\ln i(x,y)] + \text{FT}[\ln r'(x,y)] \tag{4.10}$$

对上式进行傅里叶反变换，得

$$\text{IFT}[\text{FT}[\ln f'(x,y)]] = \text{IFT}[\text{FT}[\ln i(x,y)]] + \text{IFT}[\text{FT}[\ln r'(x,y)]] \tag{4.11}$$

式中，IFT[…]表示傅里叶反变换。利用傅里叶变换和反变换性质，则上式可简化为

$$\ln f'(x,y) = \ln i(x,y) + \ln r'(x,y) \tag{4.12}$$

上式取指数，得

$$f'(x,y) = i(x,y)r'(x,y) \tag{4.13}$$

4.3.2 同态低通滤波应用

图 4.17 和图 4.18 所示分别为同态低通滤波应用。图 4.17(a)、图 4.17(b) 和图 4.17(c) 分别为图 4.2(a)、图 4.2(b) 和图 4.2(c) 的傅里叶变换同态低通滤波结果，图 4.18(a)、图 4.18(b) 和图 4.18(c) 分别为图 4.2(a)、图 4.2(b) 和图 4.2(c) 的余弦变换同态低通滤波结果。图 4.17 和图 4.18 中的(d)、(e) 分别为相位在 $-\pi/2 \sim \pi/2$ 和 $0 \sim 2\pi$ 范围的包裹相位分布，图 4.17(f) 和图 4.18(f) 为连续相位分布。

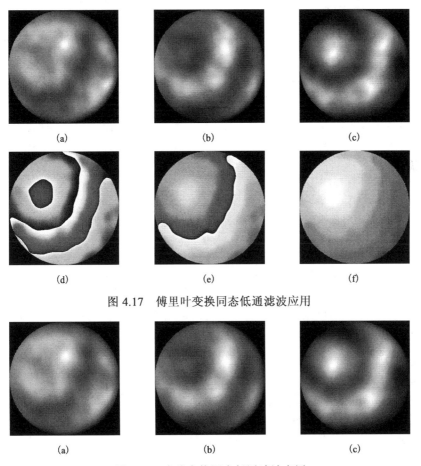

图 4.17 傅里叶变换同态低通滤波应用

图 4.18 余弦变换同态低通滤波应用

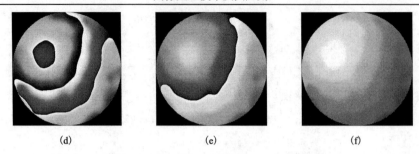

图 4.18 余弦变换同态低通滤波应用(续)

4.4 小波低通滤波

4.4.1 小波低通滤波原理

小波变换具有特征提取和低通滤波的综合作用，因而广泛应用于图像降噪。小波降噪就是对小波分解系数进行处理，然后进行图像重构，以便抑制或消除噪声。

小波降噪有多种方法，但以阈值降噪应用最为广泛。阈值降噪就是对小波分解后的大于(或小于)阈值的小波分解系数进行处理，然后利用处理后的小波系数重构降噪后的图像。阈值降噪的步骤可以归纳如下：

(1) 小波分解，即选取合适的小波函数和分解层次对图像进行小波变换，获取小波分解系数。

(2) 阈值处理，即选择合理的阈值，对大于(或小于)阈值的小波分解系数进行处理。

(3) 小波重构，即利用低频系数和经过处理的高频系数进行小波反变换，进而重构降噪图像。

4.4.2 小波低通滤波算法

1) 小波分解

MATLAB 利用 wavedec2 函数进行多层二维离散小波分解，其主要用法如下：

(1) [C, S] = wavedec2(X, N, 'wname')，进行矩阵 X 的 N 级小波分解，其中 wname 为小波名。

(2) [C, S] = wavedec2(X, N, Lo_D, Hi_D)，采用小波分解滤波器进行二维小波分解，其中 Lo_D 和 Hi_D 分别为进行小波分解的低通和高通滤波器。

2) 阈值计算

MATLAB 利用 wdcbm2 函数计算阈值，其主要用法如下：

[THR, NKEEP] = wdcbm2(C, S, ALPHA, M)，返回阈值 THR 和系数个数 NKEEP，其中 ALPHA 和 M 为大于 1 的实数。典型情况下进行图像压缩时取 ALPHA = 1.5，进行图像降噪时取 ALPHA = 3。通常情况下 M 的取值范围为[prod(S(1,:)) 6prod(S(1,:))]，默认时 M = prod(S(1,:))。

3) 小波重构

MATLAB 利用 wdencmp 函数进行图像降噪(或压缩),并在降噪(或压缩)后进行多层二维离散小波重构,其主要用法如下:

XC = wdencmp ('lvd', C, S, 'wname', N, THR, SORH),进行图像降噪,并在降噪后进行小波重构,其中 lvd 表示与层有关的阈值选项,SORH 选's'表示软阈值,选'h'表示硬阈值。

图 4.19 所示为小波变换低通滤波结果。图 4.19(a)为原始灰度图像,图 4.19(b)为原始灰度图像经过小波变换低通滤波后的灰度图像。

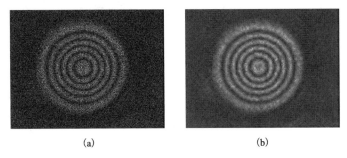

图 4.19 小波变换低通滤波结果

4.4.3 小波低通滤波应用

图 4.20 所示为小波低通滤波结果。图 4.20(a)、图 4.20(b)和图 4.20(c)分别为图 4.2(a)、图 4.2(b)和图 4.2(c)的小波低通滤波结果。图 4.20 中的(d)和(e)分别为相位在 $-\pi/2 \sim \pi/2$ 和 $0 \sim 2\pi$ 范围的包裹相位分布,图 4.20(f)为连续相位分布。

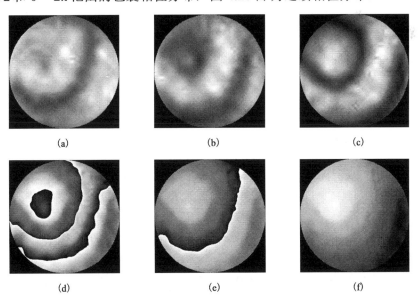

图 4.20 小波变换低通滤波结果

第 5 章 图 像 增 强

图像增强(image enhancement)是指通过突出图像有用信息或抑制图像无用信息,以提高图像质量、改善图像效果。图像增强主要包括直方图变换增强、灰度变换增强、空域滤波增强和频域滤波增强等。

5.1 直方图变换

直方图变换是指根据一定变换关系对输入像素进行灰度处理以获取输出像素。设 r_k 和 s_k 分别为输入和输出图像的第 k 级灰度值,则其直方图变换可表示为

$$s_r = \text{HT}\{r_k\} \tag{5.1}$$

式中,HT$\{\cdots\}$ 表示直方图变换,它表征输入和输出像素之间的灰度映射关系。

5.1.1 直方图

直方图(histogram)是指数字图像各级灰度(或颜色)及其出现频次之间的关系图形。横坐标表示灰度(或颜色),纵坐标表示相应灰度(或颜色)出现的频次。设灰度图像的灰度级为 L,则第 k 级灰度值可表示为 r_k,其中 $k=0,1,\cdots,L-1$。例如,8 位无符号整型灰度图像的灰度级 $L=2^8=256$,其各级灰度值可表示为 $r_0=0$(黑)、$r_1=1,\cdots,r_{255}=255$(白)。

设第 k 级灰度(或颜色)r_k 出现的频次为 $n(r_k)$,则由 $n(r_k)-r_k$ 所绘制的图形即为直方图。若所有灰度(或颜色)出现的频次总数为 n(即像素总数),则第 k 级灰度(或颜色)r_k 出现的概率可表示为

$$p(r_k) = \frac{n(r_k)}{n} \tag{5.2}$$

式中,$\sum p(r_k) = 1$。

MATLAB 利用 imhist 函数显示图像直方图,其主要用法如下:

(1) imhist(I),显示灰度图像直方图,横坐标表示灰度,纵坐标表示相应灰度出现的频次。

(2) imhist(X, map),显示索引图像直方图,横坐标表示颜色,纵坐标表示相应颜色出现的频次。

图 5.1 所示为曝光不足灰度图像及其直方图。图 5.1(a)为曝光不足灰度图像,图 5.1(b)为相应直方图。

图 5.2 所示为曝光过量灰度图像及其直方图。图 5.2(a)为曝光过量灰度图像,图 5.2(b)为相应直方图。

 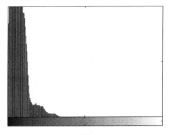

(a)　　　　　　　　　　　　　(b)

图 5.1　曝光不足图像及其直方图

 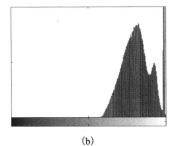

(a)　　　　　　　　　　　　　(b)

图 5.2　曝光过量图像及其直方图

图 5.3 所示为低对比度灰度图像及其直方图。图 5.3(a) 为低对比度灰度图像,图 5.3(b) 为相应直方图。

 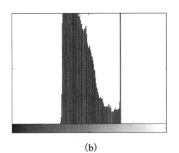

(a)　　　　　　　　　　　　　(b)

图 5.3　低对比度图像及其直方图

图 5.4 所示为高对比度灰度图像及其直方图。图 5.4(a) 为高对比度灰度图像,图 5.4(b) 为相应直方图。

 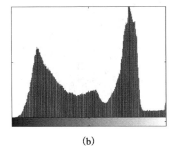

(a)　　　　　　　　　　　　　(b)

图 5.4　高对比度图像及其直方图

图 5.5 所示为索引图像及其直方图。图 5.4(a)为索引图像，图 5.4(b)为相应直方图。

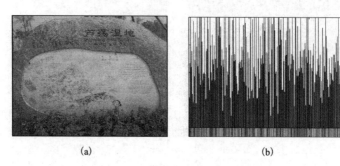

图 5.5 索引图像及其直方图

5.1.2 直方图均衡化

直方图均衡化可表示为

$$s_k = \sum_{l=0}^{k} p(r_l) = \sum_{l=0}^{k} \frac{n(r_l)}{n} \quad (k = 0, 1, \cdots, L-1) \tag{5.3}$$

式中，L 为图像灰度级，n 为所有灰度(或颜色)出现的频次总数，r_k 和 s_k 分别为输入和输出图像的第 k 级灰度，$n(r_k)$ 和 $p(r_k)$ 分别为输入图像第 k 级灰度(或颜色)出现的频次和概率。

MATLAB 利用 histeq 函数进行直方图均衡化，通过图像灰度(或索引图像颜色)变换，使输出图像直方图与指定直方图匹配，以提高图像对比度，其主要用法如下：

(1) J = histeq(I, n)，对输入图像进行变换，返回具有 n 灰度级的输出图像，使输出图像直方图变得平坦。当 n 比输入图像灰度级小得越多，则输出图像直方图越平坦。灰度级 n 的默认值是 64。

(2) J = histeq(X, map)，对索引图像颜色进行变换，使其灰度分量的直方图变得平坦，并返回变换后的颜色。

图 5.6 所示为均衡化结果。图 5.6(a)为原始图像，图 5.6(b)为均衡化图像，图 5.6(c)和图 5.6(d)分别为原始图像和均衡化图像的直方图。

图 5.6 均衡化结果

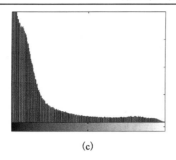
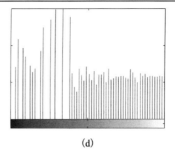

图 5.6 均衡化结果(续)

5.1.3 直方图自适应均衡化

除了可以采用直方图均衡化方法之外,还可采用直方图自适应均衡化方法增强图像对比度。直方图均衡化是全域方法,即对整个图像进行操作,而直方图自适应均衡化是局域方法,它首先把图像划分为多个子区,然后对每一子区分别采用直方图均衡化进行对比度增强,最后再通过双线性插值消除子区边界。为了避免放大图像噪声,则对图像对比度进行限制,尤其是限制均匀子区对比度。

MATLAB 利用 adaphisteq 函数进行直方图自适应均衡化,以提高图像对比度。其主要用法如下:

(1) J = adapthisteq(I),通过直方图自适应均衡化,对输入图像进行变换,提高图像对比度。

(2) J = adapthisteq(I, param1, val1, param2, val2, …),通过指定的参数及其数值对输入图像进行直方图自适应均衡化,提高图像对比度。其中参数及其数值选项如下:'NumTiles'用于指定子区数量,缺省时为[8 8];'ClipLimit'用于指定对比度,缺省时为 0.01;'NBins'用于指定灰度级,缺省时为 256;'Range'用于指定输出灰度范围,选择'original'是指输出灰度范围等于输入灰度范围,选择'full'是指输出灰度范围等于输出图像的数据类型范围,缺省是指'full';'Distribution'用于指定子区直方图形状,选择'uniform'是指平坦直方图,选择'rayleigh'是指钟形直方图,选择'exponential'是指曲线直方图,缺省是指'uniform';'Alpha'用于指定分布参数,仅对'rayleigh'或'exponential'选项适用,缺省时为 0.4。

图 5.7 所示为自适应均衡化结果。图 5.7(a)为原始图像,图 5.7(b)为自适应均衡化图像,图 5.7(c)和图 5.7(d)分别为原始图像和自适应均衡化图像的直方图。

图 5.7 自适应均衡化结果

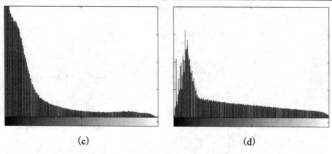

图 5.7 自适应均衡化结果(续)

5.2 灰度变换

灰度变换就是根据一定变换关系对输入像素进行灰度处理以获取输出像素。设 $f(x,y)$ 和 $g(x,y)$ 分别为输入和输出图像，则其灰度变换可表示为

$$g(x,y) = \mathrm{GT}\{f(x,y)\} \tag{5.4}$$

式中，$\mathrm{GT}\{\cdots\}$ 表示灰度变换，它表征输入和输出像素之间的灰度映射关系。

5.2.1 线性变换

若要把输入图像 $f(x,y)$ 的灰度区间 $[f_1, f_2]$ 线性变换到输出图像 $g(x,y)$ 的灰度区间 $[g_1, g_2]$，则变换后输出图像 $g(x,y)$ 可表示为

$$g(x,y) = g_1 + \frac{g_2 - g_1}{f_2 - f_1}[f(x,y) - f_1] \quad (f_1 \leqslant f(x,y) \leqslant f_2) \tag{5.5}$$

上述线性变换关系如图 5.8 所示。

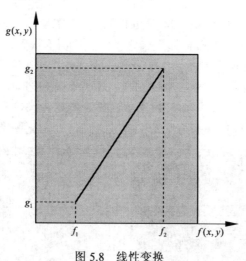

图 5.8 线性变换

为了突出有用信息，同时抑制无用信息，在图像处理中经常采用分段线性变换。以三段直线为例，则分段线性变换可表示为

$$g(x,y) = \begin{cases} g_{\min} + \dfrac{g_1 - g_{\min}}{f_1 - f_{\min}}[f(x,y) - f_{\min}] & (f_{\min} \leqslant f(x,y) < f_1) \\ g_1 + \dfrac{g_2 - g_1}{f_2 - f_1}[f(x,y) - f_1] & (f_1 \leqslant f(x,y) < f_2) \\ g_2 + \dfrac{g_{\max} - g_2}{f_{\max} - f_2}[f(x,y) - f_2] & (f_2 \leqslant f(x,y) < f_{\max}) \end{cases} \quad (5.6)$$

上述分段线性变换关系如图 5.9 所示。

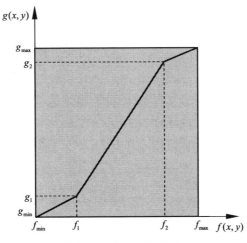

图 5.9 分段线性变换

5.2.2 非线性变换

若要把输入图像 $f(x,y)$ 的灰度区间 $[f_1, f_2]$ 非线性变换到输出图像 $g(x,y)$ 的灰度区间 $[g_1, g_2]$，则非线性变换关系可表示为

$$g(x,y) = g_1 + (g_2 - g_1)\left[\dfrac{f(x,y) - f_1}{f_2 - f_1}\right]^{\gamma} \quad (f_1 \leqslant f(x,y) \leqslant f_2,\ \gamma \neq 1) \quad (5.7)$$

变换曲线形状由 γ 确定，如图 5.10 所示。

在图像处理中也经常采用分段非线性变换。以三段曲线为例，则分段非线性变换可表示为

$$g(x,y) = \begin{cases} g_{\min} + (g_1 - g_{\min})\left[\dfrac{f(x,y) - f_{\min}}{f_1 - f_{\min}}\right]^{\gamma_1} & (f_{\min} \leqslant f(x,y) < f_1) \\ g_1 + (g_2 - g_1)\left[\dfrac{f(x,y) - f}{f_2 - f_1}\right]^{\gamma_2} & (f_1 \leqslant f(x,y) < f_2) \\ g_2 + (g_{\max} - g_2)\left[\dfrac{f(x,y) - f_2}{f_{\max} - f_2}\right]^{\gamma_3} & (f_2 \leqslant f(x,y) < f_{\max}) \end{cases} \quad (5.8)$$

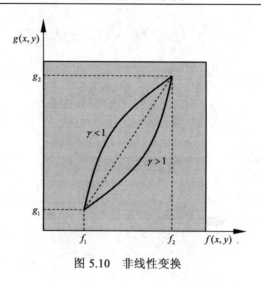

图 5.10 非线性变换

上述分段非线性变换关系如图 5.11 所示。

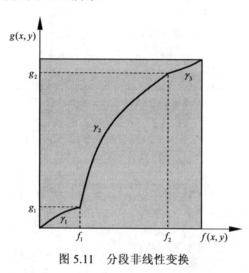

图 5.11 分段非线性变换

5.2.3 灰度变换算法

MATLAB 利用 imadjust 函数进行图像灰度变换，其主要用法如下：

(1) J = imadjust(I, [low_in; high_in], [low_out; high_out])，把输入图像灰度范围[low_in; high_in]线性映射到输出图像灰度范围[low_out; high_out]，其中 low_in, high_in, low_out, high_out 在 0~1 之间取值。低于 low_in 和高于 high_in 的数值将被截断，即低于 low_in 的数值均映射到 low_out，而高于 high_in 的数值均映射到 high_out。当空矩阵[ ]用于代替[low_in; high_in]或[low_out; high_out]时，表明采用默认值[0 1]。

(2) J = imadjust(I, [low_in; high_in], [low_out; high_out], gamma)，把输入图像映射到输出图像，其中 gamma 用于指定变换曲线的形状。gamma 小于 1 将提高中值灰度数值，

尤其是靠近低端的灰度数值。反之，gamma 大于 1 将降低中值灰度数值，尤其是靠近高端的灰度数值。当 gamma 省略时，gamma 取默认值 1（即线性变换）。

(3) J = imadjust(RGB, [low_in; high_in], [low_out; high_out], gamma)，对真彩图像的每个面（红、绿和蓝）分别进行灰度变换，把输入图像映射到输出图像。需要注意的是，当对每个面采用不同灰度范围时，则将改变原始真彩图像的颜色。

(4) J = imadjust(map, [low_in; high_in], [low_out; high_out], gamma)，对索引图像的颜色矩阵进行变换。如果 low_in, high_in, low_out, high_out, gamma 是标量，则对红、绿、蓝分量采用相同映射；如果 low_in 和 high_in、low_out 和 high_out 或 gamma 是 1×3 矢量，则对红、绿、蓝分量采用不同映射。

图 5.12 所示为灰度变换结果。图 5.12(a)为原始图像，图 5.12(b)为灰度变换图像，图 5.12(c)和图 5.12(d)分别为原始图像和灰度变换图像直方图。

图 5.12　灰度变换结果

5.3　空域滤波

空域滤波增强是指采用一定算法对输入像素邻域的灰度分布进行处理，得到输出像素的灰度分布，实现图像增强。空域滤波增强主要包括空域平滑和锐化滤波等增强技术。

5.3.1　空域平滑滤波

空域平滑滤波包括均值滤波、中值滤波和自适应滤波。这些内容在第 4 章（图像降噪）已经进行了详细讨论，在此不再重复。

5.3.2 空域锐化滤波

空域锐化滤波就是在空域对像素灰度进行差分处理，突出图像高频分量，抑制图像低频分量。空域锐化滤波主要包括梯度算子滤波和拉普拉斯算子滤波。

1. 梯度算子滤波

函数 $f(x,y)$ 在坐标 (x,y) 处的梯度可表示为

$$\nabla f = G_x \boldsymbol{i} + G_y \boldsymbol{j} = \frac{\partial f}{\partial x}\boldsymbol{i} + \frac{\partial f}{\partial y}\boldsymbol{j} \tag{5.9}$$

式中，$\nabla = \boldsymbol{i}\frac{\partial}{\partial x} + \boldsymbol{j}\frac{\partial}{\partial y}$ 为梯度算子，其中 $\boldsymbol{i},\boldsymbol{j}$ 为分别沿 x,y 方向的单位矢量。梯度幅值和方向分别为

$$|\nabla f| = \sqrt{G_x^2 + G_y^2} = \sqrt{\left(\frac{\partial f}{\partial x}\right)^2 + \left(\frac{\partial f}{\partial y}\right)^2}, \quad \theta = \arctan\left(\frac{G_y}{G_x}\right) = \arctan\left(\frac{\partial f}{\partial y}\bigg/\frac{\partial f}{\partial x}\right) \tag{5.10}$$

对数字图像处理，沿 x,y 方向的梯度分量 G_x,G_y 可用差分表示，即

$$G_x = f(x,y) - f(x+1,y), \quad G_y = f(x,y) - f(x,y+1) \tag{5.11}$$

上式也可用梯度算子表示为

$$G_x = \begin{bmatrix} 1 \\ -1 \end{bmatrix}, \quad G_y = \begin{bmatrix} 1 & -1 \end{bmatrix} \tag{5.12}$$

沿 $x'(-45°), y'(45°)$ 方向的梯度分量可用差分表示为

$$G_{x'} = f(x,y+1) - f(x+1,y), \quad G_{y'} = f(x,y) - f(x+1,y+1) \tag{5.13}$$

或用梯度算子表示为

$$G_{x'} = \begin{bmatrix} 0 & 1 \\ -1 & 0 \end{bmatrix}, \quad G_{y'} = \begin{bmatrix} 1 & 0 \\ 0 & -1 \end{bmatrix} \tag{5.14}$$

Prewitt 算子和 Sobel 算子是两种常用算子。这两种算子可分别表示为

$$G_x = \begin{bmatrix} 1 & 1 & 1 \\ 0 & 0 & 0 \\ -1 & -1 & -1 \end{bmatrix}, \quad G_y = \begin{bmatrix} 1 & 0 & -1 \\ 1 & 0 & -1 \\ 1 & 0 & -1 \end{bmatrix} \tag{5.15}$$

和

$$G_x = \begin{bmatrix} 1 & 2 & 1 \\ 0 & 0 & 0 \\ -1 & -2 & -1 \end{bmatrix}, \quad G_y = \begin{bmatrix} 1 & 0 & -1 \\ 2 & 0 & -2 \\ 1 & 0 & -1 \end{bmatrix} \tag{5.16}$$

图 5.13 所示为梯度算子滤波结果。图 5.13(a) 为原始灰度图像，图 5.13(b) 为原图与

$\mathbf{h}_1 = \begin{bmatrix} 1 & 2 & 1 \\ 0 & 0 & 0 \\ -1 & -2 & -1 \end{bmatrix}$ 进行相关运算后的图像，图 5.13(c) 为原图与 $\mathbf{h}_2 = \begin{bmatrix} 1 & 0 & -1 \\ 2 & 0 & -2 \\ 1 & 0 & -1 \end{bmatrix}$ 进行相关运算后的图像。

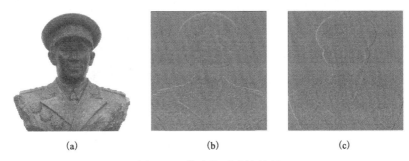

图 5.13 梯度算子滤波结果

2. 拉普拉斯算子滤波

函数 $f(x, y)$ 在坐标 (x, y) 处的拉普拉斯变换可表示为

$$\nabla^2 f = \frac{\partial^2 f}{\partial x^2} + \frac{\partial^2 f}{\partial y^2} \tag{5.17}$$

式中，$\nabla^2 = \frac{\partial^2}{\partial x^2} + \frac{\partial^2}{\partial y^2}$ 为拉普拉斯算子。拉普拉斯算子是各向同性算子。

对数字图像处理，$\frac{\partial^2 f}{\partial x^2}, \frac{\partial^2 f}{\partial y^2}$ 可用差分表示为

$$\begin{aligned} \frac{\partial^2 f}{\partial x^2} &= f(x-1, y) - 2f(x, y) + f(x+1, y) \\ \frac{\partial^2 f}{\partial y^2} &= f(x, y-1) - 2f(x, y) + f(x, y+1) \end{aligned} \tag{5.18}$$

即

$$\nabla^2 f = f(x-1, y) + f(x, y-1) - 4f(x, y) + f(x+1, y) + f(x, y+1) \tag{5.19}$$

因此，拉普拉斯算子可表示为

$$\nabla^2 = \begin{bmatrix} 0 & 1 & 0 \\ 1 & -4 & 1 \\ 0 & 1 & 0 \end{bmatrix} \tag{5.20}$$

图 5.14 所示为拉普拉斯算子滤波结果。图 5.14(a) 为原始灰度图像，图 5.14(b) 为原图与拉普拉斯模板 $\mathbf{h}_1 = \begin{bmatrix} 0 & 1 & 0 \\ 1 & -4 & 1 \\ 0 & 1 & 0 \end{bmatrix}$ 进行相关运算后的图像，图 5.14(c) 为原图与拉普拉斯

增强模板 $h_2 = \begin{bmatrix} 0 & 1 & 0 \\ 1 & -5 & 1 \\ 0 & 1 & 0 \end{bmatrix}$ 进行相关运算后的图像。

(a)　　　　　　　　　(b)　　　　　　　　　(c)

图 5.14　拉普拉斯算子滤波结果

5.4　频　域　滤　波

频域滤波增强是指将图像从空域变换到频域,在频域对图像进行滤波,再把处理结果从频域反变换到空域,进而实现图像增强。频域滤波增强主要包括频域低通滤波和频域高通滤波等。

5.4.1　频域低通滤波

频域低通滤波主要包括理想低通滤波、巴特沃斯低通滤波和指数低通滤波等。这些内容在第 4 章(图像降噪)已经进行了详细讨论,在此不再重复。

5.4.2　频域高通滤波

采用频域高通滤波可以突出图像高频信息,抑制图像低频信息。频域高通滤波主要包括理想高通滤波、巴特沃斯高通滤波和指数高通滤波等。

1. 理想高通滤波

理想高通滤波器定义为

$$H(u,v) = \begin{cases} 0 & (D(u,v) \leq D_0) \\ 1 & (D(u,v) > D_0) \end{cases} \tag{5.21}$$

式中,$H(u,v)$ 为传递函数;D_0 为截止频率;$D(u,v) = \sqrt{u^2 + v^2}$。

利用 MATLAB 语言,频域二维理想高通滤波的传递函数构造如下:
(1)离散傅里叶变换:

```
H(1:M,1:N)=ones;
for i=1:M
    for j=1:N
        if sqrt((i-(M+1)/2)^2+(j-(N+1)/2)^2)<=D
```

```
            H(i,j)=zeros;
         end
      end
end
```

(2) 离散余弦变换:

```
H(1:M,1:N)=ones;
for i=1:M
   for j=1:N
      if sqrt(i^2+j^2)<=D
         H(i,j)=zeros;
      end
   end
end
```

其中,M 和 N 分别为传递函数的行数和列数,D 为截止频率。

图 5.15 所示为理想高通滤波结果。图 5.15(a) 为原始灰度图像,图 5.15(b) 为原始灰度图像经过傅里叶变换理想高通滤波后的灰度图像,图 5.15(c) 为原始灰度图像经过余弦变换理想高通滤波后的灰度图像。

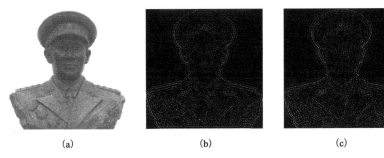

图 5.15 理想高通滤波结果

2. 巴特沃斯高通滤波

巴特沃斯高通滤波器定义为

$$H(u,v)=\frac{1}{1+\left[\dfrac{D_0}{D(u,v)}\right]^{2n}} \tag{5.22}$$

利用 MATLAB 语言,频域二维巴特沃斯(Butterworth)高通滤波的传递函数构造如下:

(1) 离散傅里叶变换:

```
for i=1:M
   for j=1:N
      H(i,j)=1/(1+(D/sqrt((i-(M+1)/2)^2+(j-(N+1)/2)^2))^(2*n));
   end
end
```

(2) 离散余弦变换：

```
for i=1:M
    for j=1:N
        H(i,j)=1/(1+(D/sqrt(i^2+j^2))^(2*n));
    end
end
```

图 5.16 所示为巴特沃斯高通滤波结果。图 5.16(a) 为原始灰度图像，图 5.16(b) 为原始灰度图像经过傅里叶变换巴特沃斯高通滤波后的灰度图像，图 5.16(c) 为原始灰度图像经过余弦变换巴特沃斯高通滤波后的灰度图像。

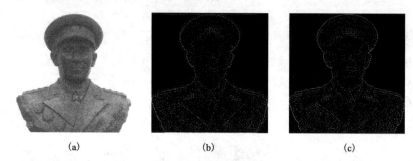

图 5.16　巴特沃斯高通滤波结果

3. 指数高通滤波

指数高通滤波器定义为

$$H(u,v) = 1 - \exp\left\{-\left[\frac{D(u,v)}{D_0}\right]^n\right\} \tag{5.23}$$

利用 MATLAB 语言，频域二维指数高通滤波的传递函数构造如下：

(1) 离散傅里叶变换：

```
for i=1:M
    for j=1:N
        H(i,j)=1-exp(-(sqrt((i-(M+1)/2)^2+(j-(N+1)/2)^2)/D)^n);
    end
end
```

(2) 离散余弦变换：

```
for i=1:M
    for j=1:N
        H(i,j)=1-exp(-(sqrt((i^2+j^2)/D)^n);
    end
end
```

图 5.17 所示为指数高通滤波结果。图 5.17(a) 为原始灰度图像,图 5.17(b) 为原始灰度图像经过傅里叶变换指数高通滤波后的灰度图像,图 5.17(c) 为原始灰度图像经过余弦变换指数高通滤波后的灰度图像。

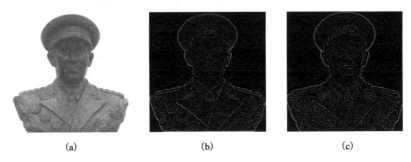

图 5.17 指数高通滤波结果

第6章 图像分割

图像分割(image segmentation)是指根据图像灰度(或颜色)在区域内部的相似性或在区域边界的差异性把图像划分为不同区域。感兴趣区域(region of interest)称为目标或对象,其余区域则称为背景。

6.1 阈值分割

阈值分割是指在图像灰度范围之内选取单个或多个阈值,然后将图像各个像素灰度与阈值进行比较,进而把图像分割为互不重叠的两个或多个区域。

6.1.1 阈值分割原理

1. 单阈值分割

单阈值分割是指把灰度小于或等于阈值的区域转换为黑,灰度值为 0;灰度大于阈值的区域转换为白,灰度值为 1。设灰度图像 $f(x,y)$ 由目标和背景组成,在目标和背景之间寻找合适的灰度作为阈值把图像分为两个区域,则分割后的图像可表示为

$$g(x,y) = \begin{cases} 0, & f(x,y) \leq T \\ 1, & f(x,y) > T \end{cases} \tag{6.1}$$

式中,T 为阈值,$g(x,y)$ 为分割后的二值图像。

MATLAB 利用 im2bw 函数根据设定的阈值把灰度、索引和真彩图像分割为二值图像,其主要用法如下:

(1) A = im2bw(I, thresh),灰度图像分割为二值图像。输入图像中灰度值大于阈值 thresh(thresh 在[0, 1]范围取值)的所有像素在输出图像中的灰度值都取 1(白),其余灰度值都取 0(黑)。

(2) A = im2bw(X, map, thresh),索引图像分割为二值图像。

(3) A = im2bw(RGB, thresh),真彩图像分割为二值图像。

图 6.1 所示为单阈值分割结果。图 6.1(a)为原始灰度图像,图 6.1(b)为原始灰度图像经过单阈值分割后的二值图像。为了提取石碑上的文字,因此文字是目标,石碑是背景。文字与石碑的灰度值不同,通过把阈值设定在文字灰度与石碑灰度之间,进而可以提取出石碑上的文字。

2. 多阈值分割

如果图像包含灰度互不相同的多个感兴趣区域,则需要进行多阈值分割。设灰度图像 $f(x,y)$ 由多个感兴趣区域组成,则分割后的图像可表示为

 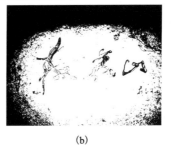

(a)   (b)

图 6.1 单阈值分割

$$g(x,y) = \begin{cases} 1, & f(x,y) \leqslant T_1 \\ 2, & T_1 < f(x,y) \leqslant T_2 \\ \vdots & \vdots \\ i, & T_{i-1} < f(x,y) \leqslant T_i \\ \vdots & \vdots \\ (n+1), & f(x,y) > T_n \end{cases} \tag{6.2}$$

式中，$T_i(i=1,2,\cdots,n)$ 为阈值，分割后图像灰度级为 $(n+1)$，对应灰度值分别为 $1,2,\cdots,n$ 和 $(n+1)$。

MATLAB 利用 imquantize 函数根据设定的多阈值把图像分割为灰度图像，其主要用法如下：

A = imquantize(I, levels)，把图像分割为灰度图像，其中 levels 为所指定的由 N 个阈值构成的矢量。输出图像包含 $(N+1)$ 个离散灰度级，灰度值分别为 $1,2,\cdots,N$ 和 $(N+1)$。

图 6.2 所示为多阈值分割结果。图 6.2(a) 为原始灰度图像，图 6.2(b) 为原始灰度图像经过多阈值分割后的灰度图像。

(a)   (b)

图 6.2 多阈值分割

6.1.2 阈值确定方法

1. 双峰法

如果图像直方图具有双峰分布，则双峰之间的谷底所对应的灰度值可取为图像分割的阈值。

图 6.3 所示为双峰法阈值分割结果。图 6.3(a)为原始灰度图像,图 6.3(b)为原始灰度图像的直方图,图 6.3(c)为经过双峰法阈值分割后的二值图像。

 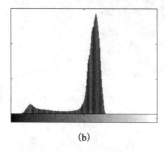 

(a) (b) (c)

图 6.3 双峰法阈值分割

2. 多峰法

如果图像直方图具有多峰分布,相邻双峰之间的谷底所对应的灰度值可取为图像分割的阈值。

图 6.4 所示为多峰法阈值分割结果。图 6.4(a)为原始灰度图像,图 6.4(b)为原始灰度图像的直方图,图 6.4(c)为经过多峰法阈值分割后的二值图像。

 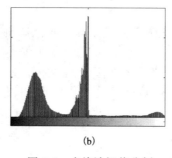 

(a) (b) (c)

图 6.4 多峰法阈值分割

3. 大津法

如果图像直方图没有明显的峰谷分布,则可采用大津法(Otsu's method)确定阈值。大津法所取阈值能够使黑白像素的类内方差最小或类间方差最大。

MATLAB 利用 graythresh 函数通过大津法计算全局阈值(global threshold),其主要用法如下:

T = graythresh(I),计算全局阈值,其中阈值已归一化,位于[0, 1]范围。利用所得阈值,灰度图像通过 im2bw 函数可分割为二值图像。

图 6.5 所示为大津法单阈值分割结果。图 6.5(a)为原始灰度图像,图 6.5(b)为原始灰度图像的直方图,图 6.5(c)为经过大津法单阈值分割后的二值图像。

MATLAB 利用 multithresh 函数通过大津法计算多级阈值(multilevel thresholds),其主要用法如下:

T = multithresh(I, N),计算多级阈值,其中 T 为包含 N 个阈值的矢量。利用所得阈值,灰度图像通过 imquantize 函数可分割为(N+1)级灰度图像。

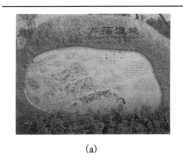 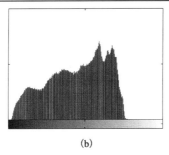 

(a) (b) (c)

图 6.5　大津法阈值分割

图 6.6 所示为大津法多阈值分割结果。图 6.6(a)为原始灰度图像，图 6.6(b)为原始灰度图像的直方图，图 6.6(c)为经过大津法多阈值分割后的灰度图像。

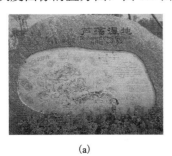 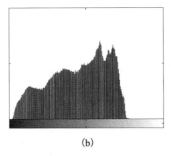 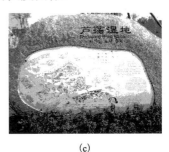

(a) (b) (c)

图 6.6　大津法多阈值分割

6.1.3　阈值分割应用

图 6.7 所示为阈值分割的应用。图 6.7(a)为原始灰度图像，图 6.7(b)为原始灰度图像的直方图，图 6.7(c)为经过阈值分割和闭合运算后的二值图像。

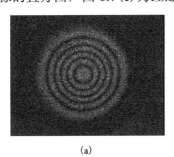 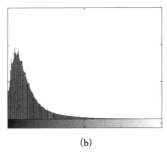 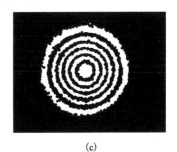

(a) (b) (c)

图 6.7　阈值分割应用

6.2　边缘检测与边界跟踪

具有不同灰度的两个区域之间存在灰度不连续，灰度不连续对应图像边缘，因此边缘是指图像上灰度不连续的点形成的轨迹。在图像边缘处，灰度一阶导数取极值，灰度二阶导数取零值。利用一阶导数极值或二阶导数零值可以检测图像边缘和跟踪图像边界。

6.2.1 边缘检测

MATLAB 利用 edge 函数识别图像边缘,寻找图像上灰度快速变化的位置。取灰度图像或二值图像作为输入,返回相同尺寸的二值图像,在返回图像中对应图像边缘的灰度为 1,其余为 0。

1. Sobel 方法

(1) BW = edge(I, 'sobel'),指定采用 Sobel 方法,缺省时采用 Sobel 方法。

(2) BW = edge(I, 'sobel', thresh),指定采用 Sobel 方法的阈值 thresh,edge 函数将忽略不比 thresh 强的边缘。如果不指定 thresh,或 thresh 取空,即[ ],那么 edge 函数自动选择阈值。

(3) BW = edge(I, 'sobel', thresh, direction),指定检测方向 direction,direction 包括水平、垂直或双向,缺省时是指双向。

(4) [BW, thresh] = edge(I, 'sobel', …),同时返回阈值 thresh。

图 6.8 所示为采用 Sobel 方法得到的图像边缘检测结果。图 6.8(a) 为原始灰度图像,图 6.8(b) 为原始灰度图像经过边缘检测后的二值图像。

图 6.8 边缘检测(Sobel 方法)

2. Prewitt 方法

(1) BW = edge(I, 'prewitt'),指定采用 Prewitt 方法,缺省时采用 Sobel 方法。

(2) BW = edge(I, 'prewitt', thresh),指定采用 Prewitt 方法的阈值 thresh,edge 函数将忽略不比 thresh 强的边缘。如果不指定 thresh,或 thresh 取空,即[ ],那么 edge 函数自动选择阈值。

(3) BW = edge(I, 'prewitt', thresh, direction),指定检测方向 direction,direction 包括水平、垂直或双向,缺省时是指双向。

(4) [BW, thresh] = edge(I, 'prewitt', …),同时返回阈值 thresh。

图 6.9 所示为采用 Prewitt 方法得到的图像边缘检测结果。图 6.9(a) 为原始灰度图像,图 6.9(b) 为原始灰度图像经过边缘检测后的二值图像。

图 6.9 边缘检测(Prewitt 方法)

3. Roberts 方法

(1) BW = edge(I, 'roberts')，指定采用 Roberts 方法，缺省时采用 Sobel 方法。

(2) BW = edge(I, 'roberts', thresh)，指定采用 Roberts 方法的阈值 thresh，edge 函数将忽略不比 thresh 强的边缘。如果不指定 thresh，或 thresh 取空，即[ ]，那么 edge 函数自动选择阈值。

(3) [BW, thresh] = edge(I, 'roberts', …)，同时返回阈值 thresh。

图 6.10 所示为采用 Roberts 方法得到的图像边缘检测结果。图 6.10(a) 为原始灰度图像，图 6.10(b) 为原始灰度图像经过边缘检测后的二值图像。

图 6.10 边缘检测(Roberts 方法)

4. Laplacian of Gaussian 方法

(1) BW = edge(I, 'log')，指定采用 Laplacian of Gaussian(LoG)方法，缺省时采用 Sobel 方法。

(2) BW = edge(I, 'log', thresh)，指定采用 LoG 方法的阈值 thresh，edge 函数将忽略不比 thresh 强的边缘。如果不指定 thresh，或 thresh 取空，即[ ]，那么 edge 函数自动选

择阈值。如果指定阈值为 0，输出图像将包含封闭等值线，因为输出图像包含了输入图像上所有零值。

(3) BW = edge(I, 'log', thresh, sigma)，指定 LoG 滤波器的标准偏差 sigma(缺省时 sigma = 2)和滤波器尺寸 n×n(n = ceil(sigma×3)×2+1)，其中 ceil(⋯)表示向正方向取整。

(4) [BW, thresh] = edge(I, 'log', ⋯)，同时返回阈值 thresh。

图 6.11 所示为采用 Laplacian of Gaussian 方法得到的图像边缘检测结果。图 6.11(a)为原始灰度图像，图 6.11(b)为原始灰度图像经过边缘检测后的二值图像。

图 6.11　边缘检测(Laplacian of Gaussian 方法)

5. Zero-Cross 方法

(1) BW = edge(I, 'zerocross', thresh, h)，采用 zero-cross 方法和滤波器 h。如果阈值 thresh 是空，即[ ]，那么 edge 函数自动选择阈值。如果指定阈值为 0，输出图像将包含封闭等值线，因为输出图像包含了输入图像上所有零值。

(2) [BW, thresh] = edge(I, 'zerocross', ⋯)，同时返回阈值 thresh。

图 6.12 所示为采用 zero-cross 方法得到的图像边缘检测结果。图 6.12(a)为原始灰度图像，图 6.12(b)为原始灰度图像经过边缘检测后的二值图像。

图 6.12　边缘检测(zero-cross 方法)

6. Canny 方法

(1) BW = edge(I, 'canny')，指定采用 Canny 方法，缺省时采用 Sobel 方法。

(2) BW = edge(I, 'canny', thresh)，指定采用 Canny 方法的阈值 thresh。其中 thresh 是两元素矢量，第一元素表示低阈值；第二元素表示高阈值。如果指定阈值 thresh 为标量，该标量值为高阈值，标量的 0.4 倍为低阈值。如果不指定 thresh，或 thresh 取空，即 [ ]，那么 edge 函数自动选择高、低阈值。

(3) BW = edge(I, 'canny', thresh, sigma)，采用 Canny 方法，指定 Gaussian 滤波器的标准偏差 sigma (缺省时 sigma = $\sqrt{2}$)，滤波器尺寸根据 sigma 自动选择。

(4) [BW, thresh] = edge(I, 'canny', …)，同时返回两元素矢量阈值 thresh。

图 6.13 所示为采用 Canny 方法得到的图像边缘检测结果。图 6.13(a) 为原始灰度图像，图 6.13(b) 为原始灰度图像经过边缘检测后的二值图像。

图 6.13　边缘检测 (Canny 方法)

6.2.2　边界跟踪

MATLAB 利用 bwboundaries 函数进行二值图像边界跟踪，其主要用法如下：

(1) B = bwboundaries(BW)，跟踪二值图像的目标外边界和目标内的孔洞边界。B 是 $P\times 1$ 数组，其中 P 是目标和孔洞数量。数组中每个单元都是 $Q\times 2$ 矩阵，其中矩阵每行都包含边界像素的行和列坐标，Q 是对应区域的边界像素数量。

(2) B = bwboundaries(BW, conn)，跟踪图像边界，其中 conn 指定边界跟踪的连通性。当 conn 等于 4 时，采用 4 连通邻域；当 conn 等于 8 或缺省时，采用 8 连通邻域。

(3) B = bwboundaries(BW, conn, options)，跟踪图像边界，其中 options 可以指定为 holes 或 noholes。当指定为 holes 或缺省时，则表示同时搜索目标和孔洞的边界；当指定为 noholes 时，则表示仅搜索目标边界。

(4) [B, L] = bwboundaries(…)，跟踪图像边界，其中 L 为表征连通区域的二维非负整数标记矩阵，数值为 0 的元素表示背景。

图 6.14 所示为图像边界跟踪结果。图 6.14(a) 为原始灰度图像，图 6.14(b) 为原始灰

度图像经过区域跟踪后的目标标记图像,图 6.14(c)为原始灰度图像经过区域跟踪后的目标和孔洞边界图像。

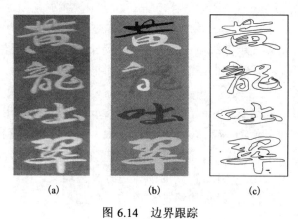

图 6.14　边界跟踪

第7章 图像恢复与再现

7.1 图像恢复

物体运动、光学离焦和大气湍流等都将引起图像模糊(blurring)或退化(degradation)。图像恢复(image restoration)就是指通过采用图像解卷积(deconvolution)算法使退化图像恢复到原始状态。

7.1.1 图像恢复原理

1. 图像退化模型

图像退化可表示为

$$g(x,y) = f(x,y) * h(x,y) + n(x,y) \tag{7.1}$$

式中,$f(x,y)$ 和 $g(x,y)$ 分别为原始图像和退化图像,*表示卷积(convolution),$n(x,y)$ 为引起图像退化的加性噪声(additive noise),$h(x,y)$ 为引起图像退化的点扩散函数(point spread function)。

利用卷积定理,式(7.1)在频域则可表示为

$$\text{FT}\{g(x,y)\} = \text{FT}\{f(x,y)\}\text{FT}\{h(x,y)\} + \text{FT}\{n(x,y)\} \tag{7.2}$$

式中,FT{⋯}表示傅里叶变换。上式也可表示为

$$G(u,v) = F(u,v)H(u,v) + N(u,v) \tag{7.3}$$

式中,$G(u,v)$ 和 $F(u,v)$ 分别为退化图像和原始图像的频谱,$N(u,v)$ 为图像噪声的频谱,$H(u,v)$ 为引起图像退化的光学传递函数(optical transfer function)。

传递函数表征线性空间不变系统在频域对脉冲的响应。光学传递函数是点扩散函数的傅里叶变换,而点扩散函数则是光学传递函数的傅里叶反变换。点扩散函数表征光学系统在空域扩展光点的程度。

2. 光学传递函数

(1) 物体运动

设图像采集期间物体在时间 τ 内沿 x 和 y 方向分别匀速移动距离 a 和 b,则传递函数可表示为

$$H(u,v) = \tau \exp[-i\pi(ua+vb)]\frac{\sin \pi(ua+vb)}{\pi(ua+vb)} \tag{7.4}$$

式中,$i = \sqrt{-1}$。

(2) 光学离焦

光学离焦的传递函数可表示为

$$H(\rho) = 2\frac{J_1(\pi a \rho)}{\pi a \rho} \tag{7.5}$$

式中，$\rho = \sqrt{u^2 + v^2}$，a 为离焦点扩散函数的直径，$J_1(\cdots)$ 为第一类一阶贝塞尔函数。

(3) 大气湍流

在大气湍流环境下的传递函数可写为

$$H(u,v) = \exp[-c(u^2 + v^2)^{5/6}] \tag{7.6}$$

式中，c 为与湍流性质有关的常数。

3. 图像恢复方法

根据式(7.1)和式(7.2)，图像恢复就是利用引起图像退化的点扩散函数使退化图像解卷积。解卷积是卷积的反向过程。

(1) 逆滤波恢复

如果采用逆滤波方法进行图像恢复，则恢复图像的频谱可表示为

$$\hat{F}(u,v) = \frac{G(u,v)}{H(u,v)} \tag{7.7}$$

式中，$\hat{F}(u,v)$ 和 $G(u,v)$ 分别为恢复图像和退化图像的频谱，$H(u,v)$ 为引起图像退化的光学传递函数。利用傅里叶反变换，则恢复图像可表示为

$$\hat{f}(x,y) = \text{IFT}\{\hat{F}(u,v)\} = \text{IFT}\left\{\frac{G(u,v)}{H(u,v)}\right\} \tag{7.8}$$

式中，$\text{IFT}\{\cdots\}$ 表示傅里叶反变换。利用 $G(u,v) = \text{FT}\{g(x,y)\}$ 和 $H(u,v) = \text{FT}\{h(x,y)\}$，则上式还可写为

$$\hat{f}(x,y) = \text{IFT}\left\{\frac{\text{FT}\{g(x,y)\}}{\text{FT}\{h(x,y)\}}\right\} \tag{7.9}$$

图 7.1 所示为逆滤波图像恢复结果。图 7.1(a) 为退化图像，图 7.1(b) 为退化图像的频谱幅值，图 7.1(c) 为恢复图像的频谱幅值，图 7.1(d) 为恢复图像。

(2) 维纳滤波恢复

如果采用维纳滤波(即最小二乘滤波)方法进行图像恢复，则恢复图像的频谱可表示为

$$\hat{F}(u,v) = \frac{|H(u,v)|^2}{|H(u,v)|^2 + S_n(u,v)/S_f(u,v)} \frac{G(u,v)}{H(u,v)} \tag{7.10}$$

式中，$S_n(u,v) = |N(u,v)|^2$ 和 $S_f(u,v) = |F(u,v)|^2$ 分别为噪声和原始图像的功率谱，$S_n(u,v)/S_f(u,v)$ 称为噪信比(即信噪比的倒数)。利用傅里叶反变换，则恢复图像可表示为

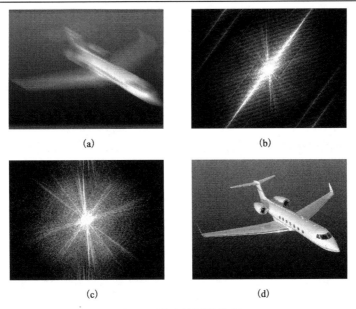

图 7.1 逆滤波图像恢复

$$\hat{f}(x,y) = \text{IFT}\left\{\frac{|\text{FT}\{h(x,y)\}|^2}{|\text{FT}\{h(x,y)\}|^2 + S_n(u,v)/S_f(u,v)} \frac{\text{FT}\{g(x,y)\}}{\text{FT}\{h(x,y)\}}\right\} \quad (7.11)$$

显然,当 $S_n(u,v)/S_f(u,v)=0$ 时,维纳滤波方法则等效于逆滤波方法。上式中的 $S_f(u,v)=|F(u,v)|^2$ 并不知道,因此在采用维纳滤波方法进行图像恢复时,通常设定 $S_n(u,v)/S_f(u,v)$ 等于常数。

(3) 约束最小二乘滤波恢复

如果采用约束最小二乘滤波方法进行图像恢复,则恢复图像的频谱可表示为

$$\hat{F}(u,v) = \frac{|H(u,v)|^2}{|H(u,v)|^2 + \lambda^{-1}|Q(u,v)|^2} \frac{G(u,v)}{H(u,v)} \quad (7.12)$$

式中,λ 为拉格朗日乘子,$Q(u,v)$ 为约束解卷积空域算子的傅里叶变换(即约束解卷积频域算子),则恢复图像可表示为

$$\hat{f}(x,y) = \text{IFT}\left\{\frac{|\text{FT}\{h(x,y)\}|^2}{|\text{FT}\{h(x,y)\}|^2 + \lambda^{-1}|Q(u,v)|^2} \frac{\text{FT}\{g(x,y)\}}{\text{FT}\{h(x,y)\}}\right\} \quad (7.13)$$

7.1.2 图像恢复算法

MATLAB 提供的图像恢复算法主要包括:维纳滤波(Wiener filter)、正则滤波(regularized filter)、Lucy-Richardson 算法(Lucy-Richardson algorithm)和盲解卷积算法(blind deconvolution algorithm)等。

1. 维纳滤波

MATLAB 利用 deconvwnr 函数采用维纳滤波通过最小二乘解(least squares solution)进行图像恢复,其主要用法如下:

(1) J = deconvwnr(I, PSF, NSR)，采用维纳滤波算法对退化图像进行解卷积，返回恢复图像。PSF 为引起图像退化的点扩散函数。NSR 为加性噪声的噪信比(noise-to-signal power ratio)，即等于信噪比的倒数，它等于 0 时等效于理想逆滤波(ideal inverse filter)。

(2) J = deconvwnr(I, PSF, NCORR, ICORR)，采用维纳滤波算法对退化图像进行解卷积，返回恢复图像。NCORR 为噪声的自相关函数(autocorrelation function)，ICORR 为退化图像的自相关函数。

图 7.2 所示为民机维纳滤波图像恢复结果。图 7.2(a)为民机退化图像，图 7.2(b)为民机恢复图像。

(a) (b)

图 7.2 民机维纳滤波图像恢复

图 7.3 所示为军机维纳滤波图像恢复结果。图 7.3(a)为军机退化图像，图 7.3(b)为军机恢复图像。

(a) (b)

图 7.3 军机维纳滤波图像恢复

2. 正则滤波

MATLAB 利用 deconvreg 函数采用正则滤波通过约束最小二乘解(constrained least squares solution)进行图像恢复，其主要用法如下：

(1) J = deconvreg(I, PSF)，采用约束最小二乘滤波算法对退化图像进行解卷积，返回恢复图像。假定退化图像是由真实图像同点扩散函数通过卷积产生，并可能含有加性噪声。

(2) J = deconvreg(I, PSF, NOISEPOWER)，采用约束最小二乘滤波算法对退化图像进行解卷积，返回恢复图像。NOISEPOWER 为加性噪声的功率，其缺省值等于 0。

(3) J = deconvreg(I, PSF, NOISEPOWER, LRANGE)，采用约束最小二乘滤波算法对退化图像进行解卷积，返回恢复图像。LRANGE 为矢量，用于指定最优解搜索范围。该算法在 LRANGE 范围内求解最优拉格朗日乘子(Lagrange multiplier)。如果 LRANGE 为标量,则该算法假定拉格朗日乘子已经给定,并等于 LRANGE。

(4) J = deconvreg(I, PSF, NOISEPOWER, LRANGE, REGOP),采用约束最小二乘滤波算法对退化图像进行解卷积,返回恢复图像。REGOP 是约束解卷积正则算子,缺省时为拉普拉斯算子(Laplacian operator)。

图 7.4 所示为民机正则滤波图像恢复结果。图 7.4(a)为民机退化图像,图 7.4(b)为民机恢复图像。

(a) (b)

图 7.4 民机正则滤波图像恢复

图 7.5 所示为军机正则滤波图像恢复结果。图 7.5(a)为军机退化图像,图 7.5(b)为军机恢复图像。

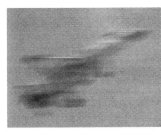 

(a) (b)

图 7.5 军机正则滤波图像恢复

3. Lucy-Richardson 算法

MATLAB 利用 deconvlucy 函数采用 Lucy-Richardson 算法在未知加性噪声的情况下通过多次迭代(multiple iteration)进行图像恢复,其主要用法如下:

(1) J = deconvlucy(I, PSF),采用 Lucy-Richardson 算法对退化图像进行恢复,其中退化图像是由点扩散函数通过卷积产生,并可能含有加性噪声。

(2) J = deconvlucy(I, PSF, NUMIT),采用 Lucy-Richardson 算法对退化图像进行恢复。NUMIT 为迭代次数,其缺省值等于 10。

(3) J = deconvlucy(I, PSF, NUMIT, DAMPAR),采用 Lucy-Richardson 算法对退化图像进行回恢复。DAMPAR 用于指定恢复图像与退化图像之间的阈值偏差,缺省值为 0(无偏差)。

(4) J = deconvlucy(I, PSF, NUMIT, DAMPAR, WEIGHT),采用 Lucy-Richardson 算法

对退化图像进行恢复。WEIGHT 用于指定分配到每个像素的权重。缺省时，WEIGHT 为单位数组，与输入图像尺寸相同。

(5) J = deconvlucy(I, PSF, NUMIT, DAMPAR, WEIGHT, READOUT)，采用 Lucy-Richardson 算法对退化图像进行恢复。READOUT 用于指定与加性噪声、相机噪声等对应的数值，缺省值为 0。

(6) J = deconvlucy(I, PSF, NUMIT, DAMPAR, WEIGHT, READOUT, SUBSMPL)，采用 Lucy-Richardson 算法对退化图像进行恢复。SUBSMPL 表示二次抽样，缺省值为 1。

图 7.6 所示为民机 Lucy-Richardson 算法图像恢复结果。图 7.6(a)为民机退化图像，图 7.6(b)为民机恢复图像。

(a)　　　　　　　　　　(b)

图 7.6　民机 Lucy-Richardson 算法图像恢复

图 7.7 所示为军机 Lucy-Richardson 算法图像恢复结果。图 7.7(a)为军机退化图像，图 7.7(b)为军机恢复图像。

(a)　　　　　　　　　　(b)

图 7.7　军机 Lucy-Richardson 算法图像恢复

4. 盲解卷积算法

MATLAB 利用 deconvblind 函数采用盲解卷积算法在未知点扩散函数的情况下通过多次迭代进行图像恢复，其主要用法如下：

(1) [J, PSF] = deconvblind(I, INITPSF)，采用最大概似算法(maximum likelihood algorithm)对退化图像进行恢复，同时返回恢复图像和恢复点扩散函数。PSF 为归一化正数组，与 INITPSF 尺寸相同。图像恢复严重依赖于 INITPSF 的初始估计尺寸，对 INITPSF 所包含的数值也较为敏感，不过通常可以指定数组 INITPSF 的数值等于 1。

(2) [J, PSF] = deconvblind(I, INITPSF, NUMIT)，采用最大概似算法(maximum

likelihood algorithm)对退化图像进行恢复,同时返回恢复图像和恢复点扩散函数。NUMIT 为迭代次数,其缺省值为 10。

(3)[J, PSF] = deconvblind(I, INITPSF, NUMIT, DAMPAR),采用最大概似算法对退化图像进行恢复,同时返回恢复图像和恢复点扩散函数。DAMPAR 用于指定恢复图像与退化图像之间的阈值偏差,缺省值为 0(无偏差)。

(4)[J, PSF] = deconvblind(I, INITPSF, NUMIT, DAMPAR, WEIGHT),采用最大概似算法对退化图像进行恢复,同时返回恢复图像和恢复点扩散函数。WEIGHT 用于指定分配到每个像素的权重。缺省时,WEIGHT 为单位数组,与输入图像尺寸相同。

(5)[J, PSF] = deconvblind(I, INITPSF, NUMIT, DAMPAR, WEIGHT, READOUT),采用最大概似算法对退化图像进行恢复,同时返回恢复图像和恢复点扩散函数。READOUT 用于指定与加性噪声、相机噪声等对应的数值,缺省值为 0。

图 7.8 所示为民机盲解卷积算法图像恢复结果。图 7.8(a)为民机退化图像,图 7.8(b)为民机恢复图像。

(a)　　　　　　　　　　　(b)

图 7.8　民机盲解卷积算法图像恢复

图 7.9 所示为军机盲解卷积算法图像恢复结果。图 7.9(a)为军机退化图像,图 7.9(b)为军机恢复图像。

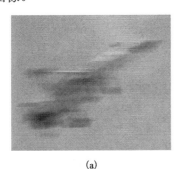 

(a)　　　　　　　　　　　(b)

图 7.9　军机盲解卷积算法图像恢复

7.1.3　图像恢复应用

图 7.10 所示为离焦图像恢复结果。图 7.10(a)为离焦图像,图 7.10(b)为恢复图像。

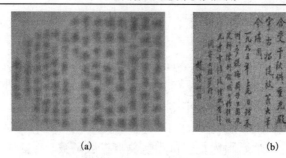

图 7.10 离焦图像恢复

图 7.11 所示为干涉条纹离焦图像恢复结果。图 7.11(a) 为干涉条纹离焦图像，图 7.11(b) 为干涉条纹恢复图像。

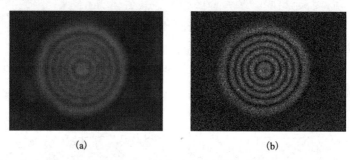

图 7.11 干涉条纹离焦图像恢复

7.2 图像再现

图像再现(image reconstruction)实际上就是典型的图像恢复过程，是图像恢复技术在光学计量领域的具体应用。

数字全息照相(digital holography)为两步成像技术，即包含记录和再现两个步骤。首先，利用物体光波和参考光波之间的干涉效应将物体光波的振幅和相位信息以干涉条纹的形式通过数码相机进行记录，并传输到计算机进行保存，其结果是一幅数字图像，称为数字全息图(digital hologram)；然后，通过再现程序模拟光波衍射效应使包含在数字全息图中的物体光波的振幅和相位信息同时得到再现，进而得到物体的立体像。

7.2.1 数字全息再现原理

设透过数字全息图的光波复振幅为 $A(x,y)$，则在菲涅耳衍射区光波复振幅分布为

$$A(u,v) = \frac{1}{\mathrm{i}\lambda z}\exp\left(\mathrm{i}\frac{2\pi}{\lambda}z\right)\exp\left[\mathrm{i}\pi\lambda z\left(\frac{u^2}{M^2\Delta x^2}+\frac{v^2}{N^2\Delta y^2}\right)\right]$$

$$\cdot \sum_{x=0}^{M-1}\sum_{y=0}^{N-1} A(x,y)\exp\left[\mathrm{i}\frac{\pi}{\lambda z}(x^2\Delta x^2 + y^2\Delta y^2)\right]\exp\left[-\mathrm{i}2\pi\left(\frac{ux}{M}+\frac{vy}{N}\right)\right] \quad (7.14)$$

$$(u=0,1,\cdots,M-1; v=0,1,\cdots,N-1)$$

式中，i=$\sqrt{-1}$ 为虚数单位，λ 为激光波长，z 为再现像到全息图的距离，$M\times N$ 为全息图尺寸，$\Delta x\times\Delta y$ 为记录全息图的 CCD 相机的像元尺寸。利用上式即可得到再现像的强度和相位分布

$$I(u,v)=|A(u,v)|^2 \tag{7.15}$$

和

$$\varphi(u,v)=\arctan\frac{\mathrm{Im}\{A(u,v)\}}{\mathrm{Re}\{A(u,v)\}} \tag{7.16}$$

式中，Re{⋯} 和 Im{⋯} 分布表示复振幅的实部和虚部。

7.2.2 数字全息再现应用

应用一：试件是正面有四个黑点的立方体，采用离轴数字全息记录系统进行数字全息照相，记录的数字全息图如图 7.12 所示。

图 7.12 数字全息图

上述数字全息图的再现结果如图 7.13 所示。图 7.13(a) 为实像平面强度分布(再现实像)；图 7.13(b) 为实像平面相位分布。

(a)   (b)

图 7.13 再现结果

应用二：试件是麻将骰子，进行离轴数字全息照相，图 7.14 所示为数字全息图。

图 7.15 所示为上述数字全息图的再现结果。图 7.15(a) 为再现实像；图 7.15(b) 为相位分布。

图 7.14 数字全息图

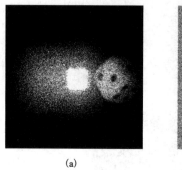 

(a) (b)

图 7.15 再现结果

第8章 图像配准与相关

8.1 图像配准

图像配准(image registration)是指不同条件下记录的两幅或两幅以上同一场景图像之间的对准过程。该过程包括指定一幅图像作为参考图像(reference image)或固定图像(fixed image)，再对其余图像，称为失配图像(misaligned image)或可动图像(moving image)，进行几何变换，以便它们与参考图像对准。几何变换可以把一幅图像的各点位置映射到另一幅图像的不同位置，因此几何变换参数的正确选择是图像配准过程的关键。

图像配准在遥感图像和医学图像等领域具有广泛应用。例如，图像配准能够用于不同条件下拍摄的遥感图像之间的配准，也可用于由不同诊疗方法得到的医学图像之间的配准。

8.1.1 图像配准原理

设参考图像由 $f_1(x,y)$ 表示，失配图像由 $f_2(x,y)$ 表示，则失配图像通过几何变换而得到的变换图像(transformed image)可表示为

$$g_2(x,y) = \text{GT}\{f_2(x,y)\} \tag{8.1}$$

式中，GT{⋯}表示几何变换(geometric transformation)。变换图像与参考图像之间的匹配程度则由相关系数(correlation coefficient)进行表征，可表示为

$$C = [g_2(x,y) - <g_2(x,y)>] \circ [f_1(x,y) - <f_1(x,y)>] \tag{8.2}$$

式中，<⋯>表示系综平均，∘表示相关运算。相关系数反映了两幅图像之间的匹配程度，当相关系数达到最大时，则表示两幅图像已经达到最佳匹配，此时变换图像即为配准图像(registered image)。

8.1.2 图像配准算法

灰度图像配准是一个迭代过程，需要确定度量标准(metric)、优化程序(optimizer)和变换类型(transformation type)。度量标准确定图像相似性，用于评价匹配程度；优化程序确定使相似性达到最小或最大的方法；变换类型确定二维变换关系，用于失配图像同参考图像进行配准。

MATLAB 利用 imregconfig 函数确定图像配准配置，其主要用法如下：

[optimizer, metric] = imregconfig(modality)，创建优化程序(optimizer)和度量标准(metric)配置，这些配置将传递给 imregister 函数进行图像配准，该函数将采用默认设置返回优化程序和度量标准基本配置。modality 用于指定图像采集方式：选择'monomodal'时，表示通过相同装置采集图像；选择'multimodal'时，为不同装置采集图像。

MATLAB 利用 imregister 函数进行图像配准，其主要用法如下：

registered = imregister(moving, fixed, type, optimizer, metric)，进行失配或可动图像变换，以便它同参考或固定图像配准。可动图像和固定图像为具有相同维数的二维图像或三维图像。type 字符串用于指定变换类型：选择'translation'时，表示平移(translation)变换；选择'rigid'时，表示平移加转动(translation and rotation)变换；选择'similarity'时，表示由平移、转动和缩放(translation, rotation, and scale)构成的非反演相似变换(nonreflective similarity transformation)；选择'affine'时，表示由平移、转动、缩放和剪切(translation, rotation, scale, and shear)构成的变换。optimizer 和 metric 分别表示优化程序和度量标准，它们由 imregconfig 函数返回。registered 为返回的配准图像(registered image)或变换图像(transformed image)。

MATLAB 利用 imfuse 函数进行图像合成或融合以显示配准结果，其主要用法如下：

(1) composite = imfuse(fixed, registered)，利用图像 fixed 和图像 registered 创建合成图像 composite。如果 fixed 和 registered 具有不同尺寸，imfuse 函数将给较小图像填 0 以便创建合成图像之前两幅图像具有相同尺寸。

(2) composite = imfuse(fixed, registered, method)，采用由参数 method 指定的算法进行图像合成。参数 method 选项如下：选择'falsecolor'或缺省时，创建合成真彩图像，以不同颜色显示重叠图像(灰色区域表示两幅图像具有相同亮度，品红和绿色区域表示亮度不同)；选择'blend'时，表示利用 alpha 混合重叠图像；选择'diff'时，表示创建差值图像；选择'montage'时，表示参考图像和配准图像并列放置而形成一幅合成图像。

8.1.3　图像配准应用

图 8.1 所示为图像配准结果。图 8.1(a) 为参考图像，图 8.1(b) 为失配图像，图 8.1(c) 为配准图像，图 8.1(d) 为合成或融合图像。

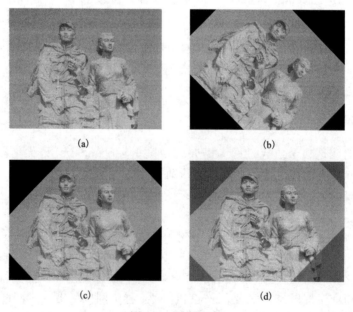

图 8.1　图像配准

8.2 图像相关

图像相关(image correlation)实际上就是典型的图像配准过程,是图像配准技术在光学计量领域的具体应用。

数字散斑相关(digital speckle correlation)采用相关模式在不显现干涉条纹的情况下可实现变形测量。

8.2.1 数字散斑相关原理

数字散斑相关通过 CCD 记录被测物体变形前后的数字散斑图(digital specklegram),对两个数字散斑图进行相关运算,找到相关系数极值点,进而得到相应的位移或变形。由于散斑分布的随机性,散斑场上的每一点周围区域(称为子区)中的散斑分布与其他点周围区域中的散斑分布互不相同,因此散斑场上以某点为中心的子区可作为该点位移和变形信息的载体,通过分析和搜索该子区的移动和变化,便可获得该点的位移和变形。

如图 8.2 所示,设对应于物体变形前后的灰度场分别用 $I(x,y)$ 和 $I'(x',y')$ 表示。在变形前的灰度场 $I(x,y)$ 中,以物体上被测点 $P(x,y)$ 为中心取子区 A,设其尺寸为 $m \times n$ 像素(通常取矩形),当发生位移或变形后灰度场 $I(x,y)$ 中的子区 A 移至灰度场 $I'(x',y')$ 中的子区 A' 位置,相应地 $P(x,y)$ 移至 $P'(x',y')$。由散斑统计特性可知,此时子区 A 与 A' 的相关系数取得极大值,因此根据相关函数的峰值就可确定子区 A' 的位置。

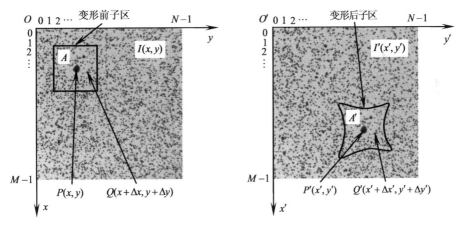

图 8.2 数字散斑相关原理

1. 位移表征

在相关计算中首先需要寻找合适变量来表征变形前后散斑图中子区的位移和变形,然后来判断变形后图像中的某个子区是否与变形前图像中给定子区相对应。设变形前后子区中心点分别为 $P(x,y)$ 和 $P'(x',y')$,如图 8.2 所示,则

$$\begin{aligned} x' &= x + u(x,y) \\ y' &= y + v(x,y) \end{aligned} \tag{8.3}$$

式中，$u(x,y)$ 和 $v(x,y)$ 分别为子区中心点在 x 和 y 方向的位移分量。

考虑变形前子区内与点 $P(x,y)$ 相邻的任意点 $Q(x+\Delta x, y+\Delta y)$，其中 Δx 和 Δy 分别为变形前子区内任意点 $Q(x+\Delta x, y+\Delta y)$ 与子区中心点 $P(x,y)$ 在 x 和 y 方向的距离。设变形前子区内的任意点 $Q(x+\Delta x, y+\Delta y)$ 在变形后移到 $Q'(x'+\Delta x', y'+\Delta y')$，则 $\Delta x'$ 和 $\Delta y'$ 可表示为

$$\begin{cases} \Delta x' = \Delta x + \Delta u(x,y) \\ \Delta y' = \Delta y + \Delta v(x,y) \end{cases} \tag{8.4}$$

式中，$\Delta u(x,y)$ 和 $\Delta v(x,y)$ 通过展开又可表示为

$$\begin{cases} \Delta u(x,y) = \dfrac{\partial u(x,y)}{\partial x}\Delta x + \dfrac{\partial u(x,y)}{\partial y}\Delta y \\ \Delta v(x,y) = \dfrac{\partial v(x,y)}{\partial x}\Delta x + \dfrac{\partial v(x,y)}{\partial y}\Delta y \end{cases} \tag{8.5}$$

综合上式，得

$$\begin{cases} x' + \Delta x' = x + u(x,y) + \left[1 + \dfrac{\partial u(x,y)}{\partial x}\right]\Delta x + \dfrac{\partial u(x,y)}{\partial y}\Delta y \\ y' + \Delta y' = y + v(x,y) + \dfrac{\partial v(x,y)}{\partial x}\Delta x + \left[1 + \dfrac{\partial v(x,y)}{\partial y}\right]\Delta y \end{cases} \tag{8.6}$$

因此，物体发生变形后，子区中心点将由 $P(x,y)$ 移到 $P'(x+u,y+v)$，子区内任意点将由 $Q(x+\Delta x,y+\Delta y)$ 移到 $Q'\left[x+u+\left(1+\dfrac{\partial u}{\partial x}\right)\Delta x+\dfrac{\partial u}{\partial y}\Delta y, y+v+\dfrac{\partial v}{\partial x}\Delta x+\left(1+\dfrac{\partial v}{\partial y}\right)\Delta y\right]$。由此可见，子区内任何点的位移都可以通过子区中心点的位移 u 和 v 及其位移导数 $\dfrac{\partial u}{\partial x}$、$\dfrac{\partial u}{\partial y}$、$\dfrac{\partial v}{\partial x}$ 和 $\dfrac{\partial v}{\partial y}$ 来表示，因此子区中心点的位移及其导数完全可以描述子区的位移和变形。

2. 相关系数

图像灰度分布是物体位移和变形信息的载体，数字散斑相关就是寻找图像局部区域（子区）灰度分布的最佳匹配程度，因此在相关分析中需要建立表示变形前后图像匹配程度的相关系数公式。相关系数公式有很多种，下面是最常用的相关系数公式：

$$C = (I_2 - <I_2>) \circ (I_1 - <I_1>) \tag{8.7}$$

式中，$<\cdots>$ 表示系综平均，$\circ$ 表示相关运算。相关系数反映两个图像子区之间的相似程度，相关系数等于 1 表示完全相关，相关系数等于 0 表示完全不相关。在数字散斑相关法中，通过求解相关系数的极大值，可实现位移和变形的提取。相关系数的峰值分布如图 8.3 所示。

3. 相关系统

数字散斑相关系统如图 8.4 所示。可以采用白光照射试件，也可以采用激光照射试件。为使试件表面光场均匀分布，可以采用对称入射方法。

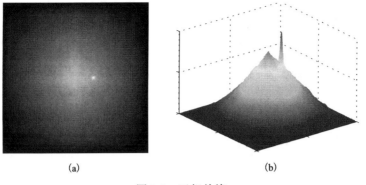

图 8.3 互相关峰

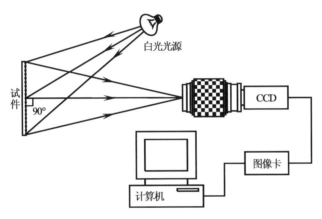

图 8.4 数字散斑相关系统

8.2.2 数字散斑相关应用

图 8.5 所示为记录的物体变形前后的白光数字散斑图。

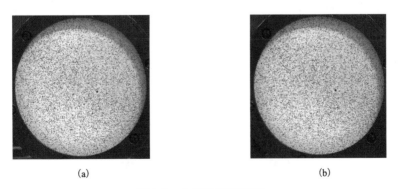

图 8.5 白光数字散斑图

图 8.6 所示为进行相关计算后得到的两个相互垂直方向的位移分量分布。图 8.6(a) 和图 8.6(c) 表示竖直方向位移分量(向下为正);图 8.6(b) 和图 8.6(d) 表示水平方向位移分量(向右为正)。图 8.6(c) 和图 8.6(d) 的水平面上同时给出了等值条纹分布。

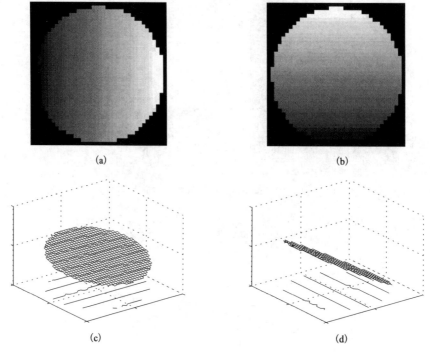

图 8.6 位移分量分布

图 8.7 所示为位移大小和方向分布。图 8.7(a) 和图 8.7(c) 表示位移大小；图 8.7(b) 和图 8.7(d) 表示位移方向。图 8.7(c) 和图 8.7(d) 的水平面上同时给出了等值条纹分布。

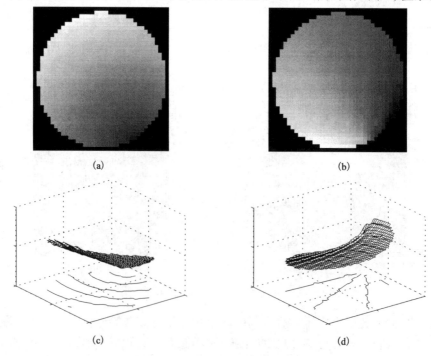

图 8.7 位移大小和方向分布

第 9 章 图像形态运算

图像形态运算(morphological operation)是指基于形状(shape)运算的图像处理技术，即输入图像通过结构元素作用而产生输出图像。在形态运算中，通过输入图像的每个像素同其邻域像素之间的比较，而获取输出图像的对应像素。通过选择邻域的尺寸和形状，可以对输入图像中的特定形状进行形态运算。

9.1 集 合 概 念

形态运算的数学基础是集合论(set theory)。集合(set)是指具有某种性质的事物的总体。集合中的单个事物称为元素(element)。

设 A 是 Z^2 空间中的集合，若 $a = (a_1, a_2)$ 是集合 A 的元素，则记作 $a \in A$（即 a 属于 A）；反之，若 $a = (a_1, a_2)$ 不是集合 A 的元素，则记作 $a \notin A$（即 a 不属于 A）。不包含任何元素的集合称为空集(null or empty set)，记作 $\varnothing$。

如果集合 A 的每个元素也是集合 B 的元素，那么集合 A 称为集合 A 的子集(subset)，记作 $A \subseteq B$（即 A 包含于 B 或 B 包含 A）。由集合 A 和 B 所有元素构成的集合 C 称为 A 和 B 的并集(union)，记作 $C = A \cup B$。由集合 A 和 B 相同元素构成的集合 D 称为 A 和 B 的交集(intersection)，记作 $D = A \cap B$。

如果集合 A 和集合 B 没有相同元素，那么 A 和 B 是无交集的(disjoint)或互斥的(mutually exclusive)，记作 $A \cap B = \varnothing$。由集合 A 以外的所有元素构成的集合称为 A 的补集(complement)，记作 A^c，即 $A^c = \{w \mid w \notin A\}$。集合 A 和 B 差值记作 $A - B$，即 $A - B = \{w \mid w \in A, w \notin B\} = A \cap B^c$。

集合 A 的反射(reflection)或反演(inversion)记为 $\hat{A}$，则 $\hat{A}$ 可表示为 $\hat{A} = \{-a \mid a \in A\}$。集合 A 的平移(translation)记为 $(A)_z$，则 $(A)_z$ 可表示为 $(A)_z = A + z = \{a + z \mid a \in A\}$，其中 $z = (x, y)$。

9.2 结 构 元 素

9.2.1 结构元素形成

结构元素(structuring element)是指仅由 0 和 1 构成的具有任意形状和尺寸的矩阵。结构元素中值为 1 的像素分布确定结构元素邻域的尺寸和形状。

二维或平面结构元素通常比需要处理的图像小很多。结构元素中心像素(即原点)即为感兴趣的像素(即需要处理的像素)。常用二维结构元素如下：

$$\begin{bmatrix} 0 & 0 & 0 & 1 & 0 & 0 & 0 \\ 0 & 0 & 1 & 1 & 1 & 0 & 0 \\ 0 & 1 & 1 & 1 & 1 & 1 & 0 \\ 1 & 1 & 1 & 1 & 1 & 1 & 1 \\ 0 & 1 & 1 & 1 & 1 & 1 & 0 \\ 0 & 0 & 1 & 1 & 1 & 0 & 0 \\ 0 & 0 & 0 & 1 & 0 & 0 & 0 \end{bmatrix} (菱形) \quad \begin{bmatrix} 0 & 0 & 0 & 1 & 0 & 0 & 0 \\ 0 & 1 & 1 & 1 & 1 & 1 & 0 \\ 0 & 1 & 1 & 1 & 1 & 1 & 0 \\ 1 & 1 & 1 & 1 & 1 & 1 & 1 \\ 0 & 1 & 1 & 1 & 1 & 1 & 0 \\ 0 & 1 & 1 & 1 & 1 & 1 & 0 \\ 0 & 0 & 0 & 0 & 0 & 0 & 0 \end{bmatrix} (盘形)$$

$$\begin{bmatrix} 0 & 0 & 1 & 1 & 1 & 0 & 0 \\ 0 & 1 & 1 & 1 & 1 & 1 & 0 \\ 1 & 1 & 1 & 1 & 1 & 1 & 1 \\ 1 & 1 & 1 & 1 & 1 & 1 & 1 \\ 1 & 1 & 1 & 1 & 1 & 1 & 1 \\ 0 & 1 & 1 & 1 & 1 & 1 & 0 \\ 0 & 0 & 1 & 1 & 1 & 0 & 0 \end{bmatrix} (八边形) \quad \begin{bmatrix} 1 & 1 & 1 & 1 & 1 & 1 \\ 1 & 1 & 1 & 1 & 1 & 1 \\ 1 & 1 & 1 & 1 & 1 & 1 \\ 1 & 1 & 1 & 1 & 1 & 1 \\ 1 & 1 & 1 & 1 & 1 & 1 \\ 1 & 1 & 1 & 1 & 1 & 1 \end{bmatrix} (正方形)$$

三维或非平面结构元素用 0 或 1 确定结构元素在 x,y 面的范围,并增加高度值确定第三维方向尺寸。常用三维结构元素的邻域及其高度如下:

$$\begin{bmatrix} 1 & 1 & 1 & 1 & 1 & 1 \\ 1 & 1 & 1 & 1 & 1 & 1 \\ 1 & 1 & 1 & 1 & 1 & 1 \\ 1 & 1 & 1 & 1 & 1 & 1 \\ 1 & 1 & 1 & 1 & 1 & 1 \\ 1 & 1 & 1 & 1 & 1 & 1 \\ 1 & 1 & 1 & 1 & 1 & 1 \end{bmatrix}, \begin{bmatrix} 0.2499 & 0.3749 & 0.4999 & 0.6248 & 0.4999 & 0.3749 & 0.2499 \\ 0.3749 & 0.4999 & 0.6248 & 0.7498 & 0.6248 & 0.4999 & 0.3749 \\ 0.4999 & 0.6248 & 0.7498 & 0.8748 & 0.7498 & 0.6248 & 0.4999 \\ 0.6248 & 0.7498 & 0.8748 & 0.9997 & 0.8748 & 0.7498 & 0.6248 \\ 0.4999 & 0.6248 & 0.7498 & 0.8748 & 0.7498 & 0.6248 & 0.4999 \\ 0.3749 & 0.4999 & 0.6248 & 0.7498 & 0.6248 & 0.4999 & 0.3749 \\ 0.2499 & 0.3749 & 0.4999 & 0.6248 & 0.4999 & 0.3749 & 0.2499 \end{bmatrix} (椭球形)$$

MATLAB 利用 strel 函数创建具有任意尺寸和形状的结构元素,其主要用法如下:

(1) SE = strel(shape, parameters),创建结构元素 SE,其中 shape 可指定为平面或非平面结构元素。对平面结构元素,shape 可选为 arbitrary、pair、diamond、periodicline、disk、rectangle、line、square 或 octagon;对非平面结构元素,shape 可选为 arbitrary 或 ball。

(2) SE = strel('arbitrary', NHOOD),创建平面结构元素。NHOOD 指定邻域,它是由 0 和 1 构成的矩阵,其中 1 定义形态运算的邻域。NHOOD 的中心(即中心元素位置)即为结构元素原点,由 floor((size(NHOOD)+1)/2) 确定。如 NHOOD 指定为 METHOD = [1 0 0; 1 0 0; 1 0 1],则表示 SE = $\begin{bmatrix} 1 & 0 & 0 \\ 1 & 0 & 0 \\ 1 & 0 & 1 \end{bmatrix}$。

(3) SE = strel('arbitrary', NHOOD, HEIGHT),创建非平面结构元素,其中 NHOOD 指定邻域,HEIGHT 指定 NHOOD 中非零元素的高度(必须是有限实数)。

(4) SE = strel('ball', R, H, N),创建非平面球形(或椭球形)结构元素,其半径(在 X-Y 平面)为 R(非负整数)、高度为 H(实数)。N 为非负偶数(如果 N 不指定,则为 8)。

(5) SE = strel('diamond', R)，创建平面菱形结构元素，其中 R 指定结构元素原点到菱形顶点之间的距离(非负整数)。如 $SE = \begin{bmatrix} 0 & 1 & 0 \\ 1 & 1 & 1 \\ 0 & 1 & 0 \end{bmatrix}$，则 R = 1。

(6) SE = strel('disk', R, N)，创建平面盘型结构元素，其中 R 指定半径(非负整数)，N 必须取 0、4、6 或 8。如果 N 不指定，则取值为 4。

(7) SE = strel('line', LEN, DEG)，创建对称于邻域中心的平面线型结构元素，其中 LEN 是元素长度，DEG 指定线的角度(自水平轴逆时针旋转)。

(8) SE = strel('octagon', R)，创建平面八边形结构元素，其中 R 指定结构元素原点到边之间的距离(必须是非负值，且为 3 的倍数)。

(9) SE = strel('pair', OFFSET)，创建仅有 2 个元素的结构元素，其中第 1 个元素位于结构元素原点，第 2 个元素的位置由矢量 OFFSET 指定(必须是整数型 2 元素矢量)。如 $SE = \begin{bmatrix} 0 & 0 & 0 \\ 0 & 1 & 0 \\ 0 & 0 & 1 \end{bmatrix}$，则 OFFSET = [1 1]。

(10) SE = strel('periodicline', P, V)，创建含有 2P+1 个元素的平面结构元素，其中 V 为包含整数行和整数列偏移量的 2 元素矢量(1 个元素位于原点，其余元素分别位于 V, $-V, 2V, -2V, \cdots, PV, -PV$)。

(11) SE = strel('rectangle', MN)，创建平面矩形结构元素，其中 MN 指定尺寸(必须为非负整型 2 元素矢量，第 1 个元素指行数，第 2 个元素指列数)。

(12) SE = strel('square', W)，创建正方形结构元素，其中宽度为 W(非负整数)。

9.2.2 结构元素分解

为了提高运算性能，strel 函数将把尺寸较大的结构元素分为小块，即结构元素分解(structuring element decomposition)。

例如，3×3 正方形结构元素不会进行分解。当输入 SE = strel('square', 3) 时，则显示：

```
SE =
Flat STREL object containing 9 neighbors.
Neighborhood:
    1   1   1
    1   1   1
    1   1   1
```

然而，4×4 正方形结构元素将会进行分解。当输入 SE = strel('square', 4) 时，则显示：

```
SE =
Flat STREL object containing 16 neighbors.
Decomposition: 2 STREL objects containing a total of 8 neighbors
Neighborhood:
    1   1   1   1
    1   1   1   1
    1   1   1   1
    1   1   1   1
```

显示表明，具有 16 个邻域的结构元素分成了只有 8 个邻域总数的 2 个结构元素。

在 MATLAB 中，通过 getsequence 函数，可观察结构元素的分解情况，如输入 D = getsequence(SE)，则显示：

$$D = $$
$$\text{2x1 array of STREL objects}$$

即 D 是 2×1 数组。再输入 D(1) 和 D(2)，则分别显示：

ans =
Flat STREL object containing 4 neighbors.
Neighborhood:
 1
 1
 1
 1

和

ans =
Flat STREL object containing 4 neighbors.
Neighborhood:
 1 1 1 1

显然，4×4 正方形结构元素 $\begin{bmatrix} 1 & 1 & 1 & 1 \\ 1 & 1 & 1 & 1 \\ 1 & 1 & 1 & 1 \\ 1 & 1 & 1 & 1 \end{bmatrix}$ 可分解为 4×1 结构元素 $\begin{bmatrix} 1 \\ 1 \\ 1 \\ 1 \end{bmatrix}$ 和 1×4 结构元素 $\begin{bmatrix} 1 & 1 & 1 & 1 \end{bmatrix}$。

盘形和球形结构元素的分解是近似的，而所有其他形状的结构元素的分解都是精确的。

9.3 膨胀和腐蚀

膨胀(dilation)和腐蚀(erosion)是最基本的形态运算。膨胀增加对象边界像素，而腐蚀则减小对象边界像素。所增加或减小的像素数量与所用结构元素的尺寸和形状有关。

9.3.1 膨胀和腐蚀原理

(1) 二值图像

设 A 是二值图像，B 是结构元素，则 A 用 B 膨胀定义为

$$A \oplus B = \{z \mid [(\hat{B})_z \cap A] \neq \varnothing\} \quad \text{或} \quad A \oplus B = \{z \mid [(\hat{B})_z \cap A] \subseteq A\} \quad (9.1)$$

A 用 B 腐蚀定义为

$$A \ominus B = \{z \mid (B)_z \subseteq A\} \quad (9.2)$$

对二值图像，膨胀和腐蚀之间存在如下对偶关系：

$$(A \oplus B)^c = A^c \ominus \hat{B}, \quad (A \ominus B)^c = A^c \oplus \hat{B} \quad (9.3)$$

显然,膨胀运算使图像扩大,而腐蚀运算则使图像收缩。

(2) 灰度图像

设 A 是灰度图像,B 是结构元素,则 A 用 B 膨胀和腐蚀分别定义为

$$(A \oplus B)(r,c) = \max\{A(r-x,c-y) + B(x,y) \mid (r-x,c-y) \in D_A, (x,y) \in D_B\} \quad (9.4)$$

$$(A \odot B)(r,c) = \min\{A(r+x,c+y) - B(x,y) \mid (r+x,c+y) \in D_A, (x,y) \in D_B\} \quad (9.5)$$

式中,max,min 分别表示取最大值和最小值,D_A, D_B 分别表示 A 和 B 的定义域。

对灰度图像,膨胀和腐蚀之间存在如下对偶关系:

$$(A \oplus B)^c(r,s) = (A^c \odot \hat{B})(r,s), \quad (A \odot B)^c(r,s) = (A^c \oplus \hat{B})(r,s) \quad (9.6)$$

式中,$A^c = -A(x,y)$,$\hat{B} = B(-x,-y)$。

9.3.2 图像边界处理

形态函数把结构元素的原点(即结构元素中心)放在输入图像上需要处理的像素位置。对图像边界像素,由结构元素确定的邻域可能超出图像边界。为了处理边界像素,形态函数给图像边界添加像素并指定像素值。针对膨胀和腐蚀,所添加像素具有不同像素值。对于膨胀运算,图像边界以外的像素用输入图像的数据类型所规定的最小值进行赋值,如二值图像和灰度图像的最小值均为 0;对于腐蚀运算,图像边界以外的像素用输入图像的数据类型所规定的最大值进行赋值,如二值图像最大值为1,灰度图像的最大值均为 255。

9.3.3 膨胀和腐蚀算法

MATLAB 利用 imdilate 函数实现膨胀运算,其主要用法如下:

(1) J = imdilate(I, SE),对二值图像和灰度图像进行膨胀,其中 SE 是指由 strel 函数返回的结构元素。对逻辑型输入图像和平面结构元素,imdilate 进行二值膨胀,否则进行灰度膨胀。

(2) J = imdilate(I, NHOOD),对二值图像和灰度图像进行膨胀,其中 NHOOD 是由 0 和 1 构成的矩阵,用于指定结构元素邻域。结构元素原点由 floor((size(NHOOD)+1)/2) 确定。

(3) J = imdilate(I, SE, SHAPE),对二值图像和灰度图像进行膨胀,其中 SHAPE 指定输出图像尺寸。SHAPE 指定为 same 或省略时,输出图像与输入图像具有相同尺寸;指定为 full 时,返回全膨胀结果。

设二值图像为 $I = \begin{bmatrix} 1 & 0 & 0 & 1 & 0 \\ 0 & 1 & 0 & 0 & 0 \\ 1 & 0 & 1 & 0 & 1 \end{bmatrix}$,结构元素为 $SE = \begin{bmatrix} 1 & 0 & 1 \\ 0 & 1 & 0 \\ 1 & 0 & 1 \end{bmatrix}$,经过 J = imdilate(I, SE) 运算,输出二值图像为 $J = \begin{bmatrix} 1 & 0 & 1 & 1 & 0 \\ 0 & 1 & 1 & 1 & 1 \\ 1 & 0 & 1 & 0 & 1 \end{bmatrix}$。

设灰度图像为 $I = \begin{bmatrix} 100 & 200 & 150 & 180 & 250 \\ 165 & 190 & 180 & 220 & 100 \\ 110 & 120 & 230 & 250 & 150 \end{bmatrix}$，结构元素为 $SE = \begin{bmatrix} 1 & 0 & 1 \\ 0 & 1 & 0 \\ 1 & 0 & 1 \end{bmatrix}$，经过 $J = \text{imdilate}(I, SE)$ 运算，输出灰度图像为 $J = \begin{bmatrix} 190 & 200 & 220 & 180 & 250 \\ 200 & 230 & 250 & 250 & 250 \\ 190 & 180 & 230 & 250 & 220 \end{bmatrix}$。

MATLAB 利用 imerode 函数实现腐蚀运算，其主要用法如下：

(1) $J = \text{imerode}(I, SE)$，对二值图像和灰度图像进行腐蚀，其中 SE 是指由 strel 函数返回的结构元素。对逻辑型输入图像和平面结构元素，imerode 进行二值腐蚀，否则进行灰度腐蚀。

(2) $J = \text{imerode}(I, NHOOD)$，对二值图像和灰度图像进行腐蚀，其中 NHOOD 是由 0 和 1 构成的矩阵，用于指定结构元素邻域。结构元素原点由 floor((size(NHOOD)+1)/2) 确定。

(3) $J = \text{imerode}(I, SE, SHAPE)$，对二值图像和灰度图像进行腐蚀，其中 SHAPE 指定输出图像尺寸。SHAPE 指定为 same 或省略时，输出图像与输入图像具有相同尺寸；指定为 full 时，返回全腐蚀结果。

设二值图像为 $I = \begin{bmatrix} 1 & 0 & 0 & 1 & 0 \\ 0 & 1 & 0 & 0 & 0 \\ 1 & 0 & 1 & 0 & 1 \end{bmatrix}$，结构元素为 $SE = \begin{bmatrix} 1 & 0 & 1 \\ 0 & 1 & 0 \\ 1 & 0 & 1 \end{bmatrix}$，经过 $J = \text{imerode}(I, SE)$ 运算，输出二值图像为 $J = \begin{bmatrix} 1 & 0 & 0 & 0 & 0 \\ 0 & 0 & 0 & 0 & 0 \\ 1 & 0 & 0 & 0 & 0 \end{bmatrix}$。

设灰度图像为 $I = \begin{bmatrix} 100 & 200 & 150 & 180 & 250 \\ 165 & 190 & 180 & 220 & 100 \\ 110 & 120 & 230 & 250 & 150 \end{bmatrix}$，结构元素为 $SE = \begin{bmatrix} 1 & 0 & 1 \\ 0 & 1 & 0 \\ 1 & 0 & 1 \end{bmatrix}$，经过 $J = \text{imerode}(I, SE)$ 运算，输出灰度图像为 $J = \begin{bmatrix} 100 & 165 & 150 & 100 & 220 \\ 120 & 100 & 120 & 150 & 100 \\ 110 & 120 & 190 & 100 & 150 \end{bmatrix}$。

9.3.4 膨胀和腐蚀应用

图 9.1 所示为二值图像膨胀和腐蚀结果。图 9.1(a) 为原始二值图像，图 9.1(b) 为膨胀后的二值图像，图 9.1(c) 为腐蚀后的二值图像。膨胀和腐蚀所用结构元素均为 $SE = \begin{bmatrix} 0 & 0 & 1 & 0 & 0 \\ 0 & 1 & 1 & 1 & 0 \\ 1 & 1 & 1 & 1 & 1 \\ 0 & 1 & 1 & 1 & 0 \\ 0 & 0 & 1 & 0 & 0 \end{bmatrix}$。

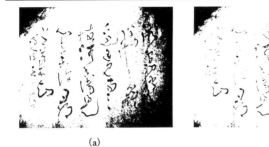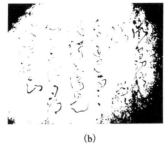

图 9.1 二值图像膨胀和腐蚀结果

图 9.2 所示为灰度图像膨胀和腐蚀结果。图 9.2(a)为原始灰度图像,图 9.2(b)为膨胀后的灰度图像,图 9.2(c)为腐蚀后的灰度图像。膨胀和腐蚀所用结构元素均为 SE = $\begin{bmatrix} 0 & 0 & 1 & 0 & 0 \\ 0 & 1 & 1 & 1 & 0 \\ 1 & 1 & 1 & 1 & 1 \\ 0 & 1 & 1 & 1 & 0 \\ 0 & 0 & 1 & 0 & 0 \end{bmatrix}$。

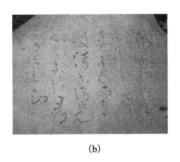

图 9.2 灰度图像膨胀和腐蚀结果

9.4 开启和闭合

9.4.1 开启和闭合原理

图像形态运算中的开启(opening)和闭合(closing)运算是由膨胀和腐蚀运算组合而成。

二值或灰度图像 A 用结构元素 B 开启定义为

$$A \circ B = (A \ominus B) \oplus B \tag{9.7}$$

上式表明,A 用 B 开启等于 A 用 B 腐蚀后再用 B 膨胀。

同理,二值或灰度图像 A 用结构元素 B 闭合定义为

$$A \bullet B = (A \oplus B) \ominus B \tag{9.8}$$

上式表明,A 用 B 闭合等于 A 用 B 膨胀后再用 B 腐蚀。

开启和闭合之间存在如下对偶关系:

$$(A \circ B)^c = A^c \bullet \hat{B}, \quad (A \bullet B)^c = A^c \circ \hat{B} \tag{9.9}$$

9.4.2 开启和闭合算法

MATLAB 利用 imopen 函数实现开启运算，其主要用法如下：

(1) J = imopen(I, SE)，用结构元素 SE 对二值图像和灰度图像进行开启。开启运算即为采用相同结构元素对输入图像先腐蚀再膨胀的二步运算。

(2) J = imopen(I, NHOOD)，用结构元素 strel(NHOOD) 对二值图像和灰度图像进行开启，其中 NHOOD 是由 0 和 1 构成的矩阵，用于指定结构元素邻域。

设二值图像为 $I = \begin{bmatrix} 1 & 0 & 0 & 1 & 0 \\ 0 & 1 & 0 & 0 & 0 \\ 1 & 0 & 1 & 0 & 1 \end{bmatrix}$，结构元素为 $SE = \begin{bmatrix} 1 & 0 & 1 \\ 0 & 1 & 0 \\ 1 & 0 & 1 \end{bmatrix}$，经过 J = imopen(I, SE)

运算，输出二值图像为 $J = \begin{bmatrix} 1 & 0 & 0 & 0 & 0 \\ 0 & 1 & 0 & 0 & 0 \\ 1 & 0 & 0 & 0 & 0 \end{bmatrix}$。

设灰度图像为 $I = \begin{bmatrix} 100 & 200 & 150 & 180 & 250 \\ 165 & 190 & 180 & 220 & 100 \\ 110 & 120 & 130 & 250 & 150 \end{bmatrix}$，结构元素为 $SE = \begin{bmatrix} 1 & 0 & 1 \\ 0 & 1 & 0 \\ 1 & 0 & 1 \end{bmatrix}$，经过

J = imopen(I, SE) 运算，输出灰度图像为 $J = \begin{bmatrix} 100 & 165 & 150 & 120 & 220 \\ 165 & 190 & 165 & 220 & 100 \\ 110 & 120 & 190 & 120 & 150 \end{bmatrix}$。

MATLAB 利用 imclose 函数实现闭合算，其主要用法如下：

(1) J = imclose(I, SE)，用结构元素 SE 对二值图像和灰度图像进行闭合。闭合运算即为采用相同结构元素对输入图像先膨胀再腐蚀的二步运算。

(2) J = imclose(I, NHOOD)，用结构元素 strel(NHOOD) 对二值图像和灰度图像进行闭合，其中 NHOOD 是由 0 和 1 构成的矩阵，用于指定结构元素邻域。

设二值图像为 $I = \begin{bmatrix} 1 & 0 & 0 & 1 & 0 \\ 0 & 1 & 0 & 0 & 0 \\ 1 & 0 & 1 & 0 & 1 \end{bmatrix}$，结构元素为 $SE = \begin{bmatrix} 1 & 0 & 1 \\ 0 & 1 & 0 \\ 1 & 0 & 1 \end{bmatrix}$，经过 J = imclose(I, SE)

运算，输出二值图像为 $J = \begin{bmatrix} 1 & 0 & 1 & 1 & 0 \\ 0 & 1 & 0 & 0 & 0 \\ 1 & 0 & 1 & 0 & 1 \end{bmatrix}$。

设灰度图像为 $I = \begin{bmatrix} 100 & 200 & 150 & 180 & 250 \\ 165 & 190 & 180 & 220 & 100 \\ 110 & 120 & 230 & 250 & 150 \end{bmatrix}$，结构元素为 $SE = \begin{bmatrix} 1 & 0 & 1 \\ 0 & 1 & 0 \\ 1 & 0 & 1 \end{bmatrix}$，经过

J = imclose(I, SE) 运算，输出灰度图像为 $J = \begin{bmatrix} 190 & 200 & 220 & 180 & 250 \\ 180 & 190 & 180 & 220 & 180 \\ 190 & 180 & 230 & 250 & 220 \end{bmatrix}$。

9.4.3 开启和闭合应用

图 9.3 所示为二值图像开启和闭合结果。图 9.3(a)为原始二值图像,图 9.3(b)为开启后的二值图像,图 9.3(c)为闭合后的二值图像。开启和闭合所用结构元素均为 SE =
$\begin{bmatrix} 0 & 0 & 1 & 0 & 0 \\ 0 & 1 & 1 & 1 & 0 \\ 1 & 1 & 1 & 1 & 1 \\ 0 & 1 & 1 & 1 & 0 \\ 0 & 0 & 1 & 0 & 0 \end{bmatrix}$。

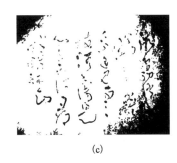

(a)   (b)   (c)

图 9.3 二值图像开启和闭合结果

图 9.4 所示为灰度图像开启和闭合结果。图 9.4(a)为原始灰度图像,图 9.4(b)为开启后的灰度图像,图 9.4(c)为闭合后的灰度图像。开启和闭合所用结构元素均为 SE =
$\begin{bmatrix} 0 & 0 & 1 & 0 & 0 \\ 0 & 1 & 1 & 1 & 0 \\ 1 & 1 & 1 & 1 & 1 \\ 0 & 1 & 1 & 1 & 0 \\ 0 & 0 & 1 & 0 & 0 \end{bmatrix}$。

(a)   (b)   (c)

图 9.4 灰度图像开启和闭合结果

附录 I 常用基本函数

I.1 Language Fundamentals

I.1.1 Entering Commands

ans	Most recent answer
clc	Clear Command Window
diary	Save Command Window text to file
format	Set display format for output
home	Send cursor home
iskeyword	Determine whether input is MATLAB keyword
more	Control paged output for Command Window

I.1.2 Matrices and Arrays

1. Array Creation and Concatenation

accumarray	Construct array with accumulation
blkdiag	Construct block diagonal matrix from input arguments
diag	Get diagonal elements or create diagonal matrix
eye	Identity matrix
false	Logical 0 (false)
freqspace	Frequency spacing for frequency response
linspace	Generate linearly spaced vectors
logspace	Generate logarithmically spaced vectors
meshgrid	Rectangular grid in 2-D and 3-D space
ndgrid	Rectangular grid in N-D space
ones	Create array of all ones
rand	Uniformly distributed pseudorandom numbers
true	Logical 1 (true)
zeros	Create array of all zeros
cat	Concatenate arrays along specified dimension
horzcat	Concatenate arrays horizontally
vertcat	Concatenate arrays vertically

2. Indexing

colon	Create vectors, array subscripting, and for-loop iterators
end	Terminate block of code, or indicate last array index
ind2sub	Subscripts from linear index
sub2ind	Convert subscripts to linear indices

3. Array Dimensions

length	Length of vector or largest array dimension
ndims	Number of array dimensions
numel	Number of array elements
size	Array dimensions
height	Number of table rows
width	Number of table variables
iscolumn	Determine whether input is column vector
isempty	Determine whether array is empty
ismatrix	Determine whether input is matrix
isrow	Determine whether input is row vector
isscalar	Determine whether input is scalar
isvector	Determine whether input is vector

4. Sorting and Reshaping Arrays

blkdiag	Construct block diagonal matrix from input argument
scircshift	Shift array circularly
ctranspose	Complex conjugate transpose
diag	Get diagonal elements or create diagonal matrix
flip	Flip order of elements
fliplr	Flip array left to right
flipud	Flip array up to down
ipermute	Inverse permute dimensions of N-D array
permute	Rearrange dimensions of N-D array
repmat	Replicate and tile array
reshape	Reshape array
rot90	Rotate array 90 degrees
shiftdim	Shift dimensions
issorted	Determine whether set elements are in sorted order
sort	Sort array elements
sortrows	Sort array rows
squeeze	Remove singleton dimensions

transpose	Transpose
vectorize	Vectorize expression

I.1.3 Operators and Elementary Operations

1. Arithmetic

plus	Addition
uplus	Unary plus
minus	Subtraction
uminus	Unary minus
times	Element-wise multiplication
rdivide	Right array division
ldivide	Left array division
power	Element-wise power
mtimes	Matrix Multiplication
mrdivide	Solve systems of linear equations $xA = B$ for x
mldivide	Solve systems of linear equations $Ax = B$ for x
mpower	Matrix power
cumprod	Cumulative product
cumsum	Cumulative sum
diff	Differences and Approximate Derivatives
prod	Product of array elements
sum	Sum of array elements
ceil	Round toward positive infinity
fix	Round toward zero
floor	Round toward negative infinity
idivide	Integer division with rounding option
mod	Modulus after division
rem	Remainder after division
round	Round to nearest integer

2. Relational Operations

<, >, <=, >=, ==, ~=	Relational operations
eq	Determine equality
ge	Determine greater than or equal to
gt	Determine greater than
le	Determine less than or equal to
lt	Determine less than
ne	Determine inequality

| isequal | Determine array equality |
| isequaln | Determine array equality, treating NaN values as equal |

3. Logical Operations

&&, ||	Logical operations with short-circuiting
and	Find logical AND
not	Find logical NOT
or	Find logical OR
xor	Logical exclusive-OR
all	Determine if all array elements are nonzero or true
any	Determine if any array elements are nonzero
false	Logical 0 (false)
find	Find indices and values of nonzero elements
islogical	Determine if input is logical array
logical	Convert numeric values to logicals
true	Logical 1 (true)

I.1.4 Special Characters

| [ ], { }, ( ), =, ', ,, ., ;, !, % | Special characters |
| colon | Create vectors, array subscripting, and for-loop iterators |

I.1.5 Data Types

1. Numeric Types

double	Convert to double precision
single	Convert to single precision
int8	Convert to 8-bit signed integer
int16	Convert to 16-bit signed integer
int32	Convert to 32-bit signed integer
int64	Convert to 64-bit signed integer
uint8	Convert to 8-bit unsigned integer
uint16	Convert to 16-bit unsigned integer
uint32	Convert to 32-bit unsigned integer
uint64	Convert to 64-bit unsigned integer
cast	Cast variable to different data type
typecast	Convert data types without changing underlying data
isinteger	Determine if input is integer array
isfloat	Determine if input is floating-point array
isnumeric	Determine if input is numeric array

isreal	Determine if array is real
isfinite	Array elements that are finite
isinf	Array elements that are infinite
isnan	Array elements that are NaN
eps	Floating-point relative accuracy
flintmax	Largest consecutive integer in floating-point format
Inf	Infinity
intmax	Largest value of specified integer type
intmin	Smallest value of specified integer type
NaN	Not-a-Number
realmax	Largest positive floating-point number
realmin	Smallest positive normalized floating-point number

2. Characters and Strings

1) Create and Concatenate Strings

blanks	Create string of blank characters
cellstr	Create cell array of strings from character array
char	Convert to character array（string）
iscellstr	Determine whether input is cell array of strings
ischar	Determine whether item is character array
sprintf	Format data into string
strcat	Concatenate strings horizontally
strjoin	Join strings in cell array into single string

2) Parse Strings

ischar	Determine whether item is character array
isletter	Array elements that are alphabetic letters
isspace	Array elements that are space characters
isstrprop	Determine whether string is of specified category
sscanf	Read formatted data from string
strfind	Find one string within another
strrep	Find and replace substring
strsplit	Split string at specified delimiter
strtok	Selected parts of string
validatestring	Check validity of text string
symvar	Determine symbolic variables in expression
regexp	Match regular expression（case sensitive）
regexpi	Match regular expression（case insensitive）
regexprep	Replace string using regular expression

regexptranslate	Translate string into regular expression

3) Compare Strings

strcmp	Compare strings with case sensitivity
strcmpi	Compare strings (case insensitive)
strncmp	Compare first n characters of strings (case sensitive)
strncmpi	Compare first n characters of strings (case insensitive)

4) Change String Case, Blanks, and Justification

blanks	Create string of blank characters
deblank	Strip trailing blanks from end of string
strtrim	Remove leading and trailing white space from string
lower	Convert string to lowercase
upper	Convert string to uppercase
strjust	Justify character array

3. Categorical Arrays

categorical	Create categorical array
iscategorical	Determine whether input is categorical array
categories	Categories of categorical array
iscategory	Test for categorical array categories
isordinal	Determine whether input is ordinal categorical array
isprotected	Determine whether categories of categorical array are protected
addcats	Add categories to categorical array
mergecats	Merge categories in categorical array
removecats	Remove categories from categorical array
renamecats	Rename categories in categorical array
reordercats	Reorder categories in categorical array
summary	Print summary of table or categorical array
countcats	Count occurrences of categorical array elements by category
isundefined	Find undefined elements in categorical array

4. Tables

table	Create table from workspace variables
array2table	Convert homogeneous array to table
cell2table	Convert cell array to table
struct2table	Convert structure array to table
table2array	Convert table to homogenous array
table2cell	Convert table to cell array
table2struct	Convert table to structure array

readtable	Create table from file
writetable	Write table to file
istable	Determine whether input is table
height	Number of table rows
width	Number of table variables
summary	Print summary of table or categorical array
intersect	Set intersection of two arrays
ismember	Array elements that are members of set array
setdiff	Set difference of two arrays
setxor	Set exclusive OR of two arrays
unique	Unique values in array
union	Set union of two arrays
join	Merge two tables by matching up rows using key variables
innerjoin	Inner join between two tables
outerjoin	Outer join between two tables
sortrows	Sort array rows
stack	Stack data from multiple variables into single variable
unstack	Unstack data from single variable into multiple variables
ismissing	Find table elements with missing values
standardizeMissing	Insert missing value indicators into table
varfun	Apply function to table variables
rowfun	Apply function to table rows

5. Structures

struct	Create structure array
fieldnames	Field names of structure, or public fields of object
getfield	Field of structure array
isfield	Determine whether input is structure array field
isstruct	Determine whether input is structure array
orderfields	Order fields of structure array
rmfield	Remove fields from structure
setfield	Assign values to structure array field
arrayfun	Apply function to each element of array
structfun	Apply function to each field of scalar structure
table2struct	Convert table to structure array
struct2table	Convert structure array to table
cell2struct	Convert cell array to structure array
struct2cell	Convert structure to cell array

6. Cell Arrays

cell	Create cell array
cell2mat	Convert cell array to numeric array
cell2struct	Convert cell array to structure array
cell2table	Convert cell array to table
celldisp	Display cell array contents
cellfun	Apply function to each cell in cell array
cellplot	Graphically display structure of cell array
cellstr	Create cell array of strings from character array
iscell	Determine whether input is cell array
iscellstr	Determine whether input is cell array of strings
mat2cell	Convert array to cell array with potentially different sized cells
num2cell	Convert array to cell array with consistently sized cells
strjoin	Join strings in cell array into single string
strsplit	Split string at specified delimiter
struct2cell	Convert structure to cell array
table2cell	Convert table to cell array

7. Function Handles

function_handle (@)	Handle used in calling functions indirectly
feval	Evaluate function
func2str	Construct function name string from function handle
str2func	Construct function handle from function name string
localfunctions	Function handles to all local functions in MATLAB file
functions	Information about function handle

8. Map Containers

containers.Map	Map values to unique keys
isKey	Determine if containers.Map object contains key
keys	Identify keys of containers.Map object
remove	Remove key-value pairs from containers.Map object
values	Identify values in containers.Map object

9. Time Series

1) Time Series Basics

append	Concatenate time series objects in time dimension
get	Query timeseries object property values
getdatasamplesize	Size of data sample in timeseries object

getqualitydesc	Data quality descriptions
getsamples	Subset of time series samples using subscripted index array
plot	Plot time series
set	Set properties of timeseries object
tsdata.event	Construct event object for timeseries object
timeseries	Create timeseries object

2) Data Manipulation

addsample	Add data sample to timeseries object
ctranspose	Transpose timeseries object
delsample	Remove sample from timeseries object
detrend	Subtract mean or best-fit line and all NaNs from timeseries object
filter	Shape frequency content of time-series
getabstime	Extract date-string time vector into cell array
getinterpmethod	Interpolation method for timeseries object
getsampleusingtime	Extract data samples into new timeseries object
idealfilter	Apply ideal (noncausal) filter to timeseries object
resample	Select or interpolate timeseries data using new time vector
setabstime	Set times of timeseries object as date strings
setinterpmethod	Set default interpolation method for timeseries object
synchronize	Synchronize and resample two timeseries objects using common time vector
transpose	Transpose timeseries object

3) Event Data

addevent	Add event to timeseries object
delevent	Remove tsdata.event objects from timeseries object
gettsafteratevent	New timeseries object with samples occurring at or after event
gettsafterevent	New timeseries object with samples occurring after event
gettsatevent	New timeseries object with samples occurring at event
gettsbeforeatevent	New timeseries object with samples occurring before or at event
gettsbeforeevent	New timeseries object with samples occurring before event
gettsbetweenevents	New timeseries object with samples occurring between events

4) Descriptive Statistics

iqr	Interquartile range of timeseries data

max	Maximum value of timeseries data
mean	Mean value of timeseries data
median	Median value of timeseries data
min	Minimum value of timeseries data
std	Standard deviation of timeseries data
sum	Sum of timeseries data
var	Variance of timeseries data

5) Time Series Collections

get (tscollection)	Query tscollection object property values
isempty (tscollection)	Determine whether tscollection object is empty
length (tscollection)	Length of time vector
plot	Plot time series
set (tscollection)	Set properties of tscollection object
size (tscollection)	Size of tscollection object
tscollection	Create tscollection object
addsampletocollection	Add sample to tscollection object
addts	Add timeseries object to tscollection object
delsamplefromcollection	Remove sample from tscollection object
getabstime (tscollection)	Extract date-string time vector into cell array
getsampleusingtime (tscollection)	Extract data samples into new tscollection object
gettimeseriesnames	Cell array of names of timeseries objects in tscollection object
horzcat (tscollection)	Horizontal concatenation for tscollection objects
removets	Remove timeseries objects from tscollection object
resample (tscollection)	Select or interpolate data in tscollection using new time vector
setabstime (tscollection)	Set times of tscollection object as date strings
settimeseriesnames	Change name of timeseries object in tscollection
vertcat (tscollection)	Vertical concatenation for tscollection objects

10. Data Type Identification

is*	Detect state
isa	Determine if input is object of specified class
iscategorical	Determine whether input is categorical array
iscell	Determine whether input is cell array
iscellstr	Determine whether input is cell array of strings
ischar	Determine whether item is character array
isfield	Determine whether input is structure array field

isfloat	Determine if input is floating-point array
ishghandle	True for Handle Graphics object handles
isinteger	Determine if input is integer array
isjava	Determine if input is Java object
islogical	Determine if input is logical array
isnumeric	Determine if input is numeric array
isobject	Determine if input is MATLAB object
isreal	Determine if array is real
isscalar	Determine whether input is scalar
isstr	Determine whether input is character array
isstruct	Determine whether input is structure array
istable	Determine whether input is table
isvector	Determine whether input is vector
class	Determine class of object
validateattributes	Check validity of array
whos	List variables in workspace, with sizes and types

11. Data Type Conversion

char	Convert to character array（string）
int2str	Convert integer to string
mat2str	Convert matrix to string
num2str	Convert number to string
str2double	Convert string to double-precision value
str2num	Convert string to number
native2unicode	Convert numeric bytes to Unicode character representation
unicode2native	Convert Unicode character representation to numeric bytes
base2dec	Convert base N number string to decimal number
bin2dec	Convert binary number string to decimal number
dec2base	Convert decimal to base N number in string
dec2bin	Convert decimal to binary number in string
dec2hex	Convert decimal to hexadecimal number in string
hex2dec	Convert hexadecimal number string to decimal number
hex2num	Convert hexadecimal number string to double-precision number
num2hex	Convert singles and doubles to IEEE hexadecimal strings
table2array	Convert table to homogenous array
table2cell	Convert table to cell array
table2struct	Convert table to structure array

array2table	Convert homogeneous array to table
cell2table	Convert cell array to table
struct2table	Convert structure array to table
cell2mat	Convert cell array to numeric array
cell2struct	Convert cell array to structure array
cellstr	Create cell array of strings from character array
mat2cell	Convert array to cell array with potentially different sized cells
num2cell	Convert array to cell array with consistently sized cells
struct2cell	Convert structure to cell array

I.1.6 Dates and Time

datenum	Convert date and time to serial date number
datevec	Convert date and time to vector of components
datestr	Convert date and time to string format
now	Current date and time as serial date number
clock	Current date and time as date vector
date	Current date string
calendar	Calendar for specified month
eomday	Last day of month
weekday	Day of week
addtodate	Modify date number by field
etime	Time elapsed between date vectors

I.2 Mathematics

I.2.1 Elementary Math

1. Arithmetic

plus	Addition
uplus	Unary plus
minus	Subtraction
uminus	Unary minus
times	Element-wise multiplication
rdivide	Right array division
ldivide	Left array division
power	Element-wise power

mtimes	Matrix Multiplication
mrdivide	Solve systems of linear equations xA = B for x
mldivide	Solve systems of linear equations Ax = B for x
mpower	Matrix power
cumprod	Cumulative product
cumsum	Cumulative sum
diff	Differences and Approximate Derivatives
prod	Product of array elements
sum	Sum of array elements
ceil	Round toward positive infinity
fix	Round toward zero
floor	Round toward negative infinity
idivide	Integer division with rounding option
mod	Modulus after division
rem	Remainder after division
round	Round to nearest integer

2. Trigonometry

sin	Sine of argument in radians
sind	Sine of argument in degrees
asin	Inverse sine in radians
asind	Inverse sine in degrees
sinh	Hyperbolic sine of argument in radians
asinh	Inverse hyperbolic sine
cos	Cosine of argument in radians
cosd	Cosine of argument in degrees
acos	Inverse cosine in radians
acosd	Inverse cosine in degrees
cosh	Hyperbolic cosine
acosh	Inverse hyperbolic cosine
tan	Tangent of argument in radians
tand	Tangent of argument in degrees
atan	Inverse tangent in radians
atand	Inverse tangent in degrees
atan2	Four-quadrant inverse tangent
atan2d	Four-quadrant inverse tangent in degrees
tanh	Hyperbolic tangent
atanh	Inverse hyperbolic tangent

csc	Cosecant of input angle in radians
cscd	Cosecant of argument in degrees
acsc	Inverse cosecant in radians
acscd	Inverse cosecant in degrees
csch	Hyperbolic cosecant
acsch	Inverse hyperbolic cosecant
sec	Secant of angle in radians
secd	Secant of argument in degrees
asec	Inverse secant in radians
asecd	Inverse secant in degrees
sech	Hyperbolic secant
asech	Inverse hyperbolic secant
cot	Cotangent of angle in radians
cotd	Cotangent of argument in degrees
acot	Inverse cotangent in radians
acotd	Inverse cotangent in degrees
coth	Hyperbolic cotangent
acoth	Inverse hyperbolic cotangent
hypot	Square root of sum of squares

3. Exponents and Logarithms

exp	Exponential
expm1	Compute exp(x)-1 accurately for small values of x
log	Natural logarithm
log10	Common (base 10) logarithm
log1p	Compute log(1+x) accurately for small values of x
log2	Base 2 logarithm and dissect floating-point numbers into exponent and mantissa
nextpow2	Exponent of next higher power of 2
nthroot	Real nth root of real numbers
pow2	Base 2 power and scale floating-point numbers
reallog	Natural logarithm for nonnegative real arrays
realpow	Array power for real-only output
realsqrt	Square root for nonnegative real arrays
sqrt	Square root

4. Complex Numbers

abs	Absolute value and complex magnitude
angle	Phase angle

complex	Create complex array
conj	Complex conjugate
cplxpair	Sort complex numbers into complex conjugate pairs
i	Imaginary unit
imag	Imaginary part of complex number
isreal	Determine if array is real
j	Imaginary unit
real	Real part of complex number
sign	Signum function
unwrap	Correct phase angles to produce smoother phase plots

5. Discrete Math

factor	Prime factors
factorial	Factorial of input
gcd	Greatest common divisor
isprime	Determine which array elements are prime
lcm	Least common multiple
nchoosek	Binomial coefficient or all combinations
perms	All possible permutations
primes	Prime numbers less than or equal to input value
rat	Rational fraction approximation
rats	Rational output

6. Polynomials

poly	Polynomial with specified roots
polyder	Polynomial derivative
polyeig	Polynomial eigenvalue problem
polyfit	Polynomial curve fitting
polyint	Integrate polynomial analytically
polyval	Polynomial evaluation
polyvalm	Matrix polynomial evaluation
residue	Convert between partial fraction expansion and polynomial coefficients
roots	Polynomial roots

7. Special Functions

airy	Airy Functions
besselh	Bessel function of third kind (Hankel function)
besseli	Modified Bessel function of first kind

besselj	Bessel function of first kind
besselk	Modified Bessel function of second kind
bessely	Bessel function of second kind
beta	Beta function
betainc	Incomplete beta function
betaincinv	Beta inverse cumulative distribution function
betaln	Logarithm of beta function
ellipj	Jacobi elliptic functions
ellipke	Complete elliptic integrals of first and second kind
erf	Error function
erfc	Complementary error function
erfcinv	Inverse complementary error function
erfcx	Scaled complementary error function
erfinv	Inverse error function
expint	Exponential integral
gamma	Gamma function
gammainc	Incomplete gamma function
gammaincinv	Inverse incomplete gamma function
gammaln	Logarithm of gamma function
legendre	Associated Legendre functions
psi	Psi (polygamma) function

8. Cartesian Coordinate System Conversion

cart2pol	Transform Cartesian coordinates to polar or cylindrical
cart2sph	Transform Cartesian coordinates to spherical
pol2cart	Transform polar or cylindrical coordinates to Cartesian
sph2cart	Transform spherical coordinates to Cartesian

9. Constants and Test Matrices

eps	Floating-point relative accuracy
flintmax	Largest consecutive integer in floating-point format
i	Imaginary unit
j	Imaginary unit
Inf	Infinity
pi	Ratio of circle's circumference to its diameter
NaN	Not-a-Number
isfinite	Array elements that are finite
isinf	Array elements that are infinite
isnan	Array elements that are NaN

company	Companion matrix
gallery	Test matrices
hadamard	Hadamard matrix
hankel	Hankel matrix
hilb	Hilbert matrix
invhilb	Inverse of Hilbert matrix
magic	Magic square
pascal	Pascal matrix
rosser	Classic symmetric eigenvalue test problem
toeplitz	Toeplitz matrix
vander	Vandermonde matrix
wilkinson	Wilkinson's eigenvalue test matrix

I.2.2　Linear Algebra

1. Matrix Operations

cross	Cross product
dot	Dot product
kron	Kronecker tensor product
surfnorm	Compute and display 3-D surface normals
tril	Lower triangular part of matrix
triu	Upper triangular part of matrix
transpose	Transpose

2. Linear Equations

cond	Condition number with respect to inversion
condest	1-norm condition number estimate
funm	Evaluate general matrix function
inv	Matrix inverse
linsolve	Solve linear system of equations
lscov	Least-squares solution in presence of known covariance
lsqnonneg	Solve nonnegative least-squares constraints problem
pinv	Moore-Penrose pseudoinverse of matrix
rcond	Reciprocal condition number
sylvester	Solve Sylvester equation $AX + XB = C$ for X
mldivide	Solve systems of linear equations $Ax = B$ for x
mrdivide	Solve systems of linear equations $xA = B$ for x

3. Matrix Decomposition

chol	Cholesky factorization
ichol	Incomplete Cholesky factorization
cholupdate	Rank 1 update to Cholesky factorization
ilu	Sparse incomplete LU factorization
lu	LU matrix factorization
qr	Orthogonal-triangular decomposition
qrdelete	Remove column or row from QR factorization
qrinsert	Insert column or row into QR factorization
qrupdate	Rank 1 update to QR factorization
planerot	Givens plane rotation
ldl	Block LDL' factorization for Hermitian indefinite matrices
cdf2rdf	Convert complex diagonal form to real block diagonal form
rsf2csf	Convert real Schur form to complex Schur form
gsvd	Generalized singular value decomposition
svd	Singular value decomposition

4. Eigenvalues and Singular Values

balance	Diagonal scaling to improve eigenvalue accuracy
cdf2rdf	Convert complex diagonal form to real block diagonal form
condeig	Condition number with respect to eigenvalues
eig	Eigenvalues and eigenvectors
eigs	Largest eigenvalues and eigenvectors of matrix
gsvd	Generalized singular value decomposition
hess	Hessenberg form of matrix
ordeig	Eigenvalues of quasitriangular matrices
ordqz	Reorder eigenvalues in QZ factorization
ordschur	Reorder eigenvalues in Schur factorization
poly	Polynomial with specified roots
polyeig	Polynomial eigenvalue problem
qz	QZ factorization for generalized eigenvalues
rsf2csf	Convert real Schur form to complex Schur form
schur	Schur decomposition
sqrtm	Matrix square root
ss2tf	Convert state-space filter parameters to transfer function form
svd	Singular value decomposition
svds	Find singular values and vectors

5. Matrix Analysis

bandwidth	Lower and upper matrix bandwidth
cond	Condition number with respect to inversion
condeig	Condition number with respect to eigenvalues
det	Matrix determinant
isbanded	Determine if matrix is within specific bandwidth
isdiag	Determine if matrix is diagonal
ishermitian	Determine if matrix is Hermitian or skew-Hermitian
issymmetric	Determine if matrix is symmetric or skew-symmetric
istril	Determine if matrix is lower triangular
istriu	Determine if matrix is upper triangular
norm	Vector and matrix norms
normest	2-norm estimate
null	Null space
orth	Orthonormal basis for range of matrix
rank	Rank of matrix
rcond	Reciprocal condition number
rref	Reduced row echelon form
subspace	Angle between two subspaces
trace	Sum of diagonal elements

6. Matrix Functions

expm	Matrix exponential
logm	Matrix logarithm
sqrtm	Matrix square root
bsxfun	Apply element-by-element binary operation to two arrays with singleton expansion enabled
arrayfun	Apply function to each element of array
accumarray	Construct array with accumulation
mpower	Matrix power

I.2.3 Statistics and Random Numbers

1. Descriptive Statistics

corrcoef	Correlation coefficients
cov	Covariance matrix
max	Largest elements in array
mean	Average or mean value of array

median	Median value of array
min	Smallest elements in array
mode	Most frequent values in array
std	Standard deviation
var	Variance

2. Random Number Generation

rand	Uniformly distributed pseudorandom numbers
randn	Normally distributed pseudorandom numbers
randi	Uniformly distributed pseudorandom integers
randperm	Random permutation
rng	Control random number generation
RandStream	Random number stream

I.2.4 Interpolation

1. 1-D Interpolation

interp1	1-D data interpolation (table lookup)
griddedInterpolant	Gridded data interpolation
pchip	Piecewise Cubic Hermite Interpolating Polynomial (PCHIP)
spline	Cubic spline data interpolation
ppval	Evaluate piecewise polynomial
mkpp	Make piecewise polynomial
unmkpp	Piecewise polynomial details
padecoef	Padé approximation of time delays
interpft	1-D interpolation using FFT method

2. Gridded Data Interpolation

interp2	Interpolation for 2-D gridded data in meshgrid format
interp3	Interpolation for 3-D gridded data in meshgrid format
interpn	Interpolation for 1-D, 2-D, 3-D, and N-D gridded data in ndgrid format
griddedInterpolant	Gridded data interpolation
ndgrid	Rectangular grid in N-D space
meshgrid	Rectangular grid in 2-D and 3-D space

3. Scattered Data Interpolation

griddata	Interpolate scattered data
griddatan	Data gridding and hypersurface fitting (dimension$\geqslant$2)

scatteredInterpolant	Scattered data interpolation

I.2.5 Fourier Analysis and Filtering

abs	Absolute value and complex magnitude
angle	Phase angle
cplxpair	Sort complex numbers into complex conjugate pairs
fft	Fast Fourier transform
fft2	2-D fast Fourier transform
fftn	N-D fast Fourier transform
fftshift	Shift zero-frequency component to center of spectrum
fftw	Interface to FFTW library run-time algorithm tuning control
ifft	Inverse fast Fourier transform
ifft2	2-D inverse fast Fourier transform
ifftn	N-D inverse fast Fourier transform
ifftshift	Inverse FFT shift
nextpow2	Exponent of next higher power of 2
unwrap	Correct phase angles to produce smoother phase plots
conv	Convolution and polynomial multiplication
conv2	2-D convolution
convn	N-D convolution
deconv	Deconvolution and polynomial division
detrend	Remove linear trends
filter	1-D digital filter
filter2	2-D digital filter

I.3 Graphics

I.3.1 2-D and 3-D Plots

1. Line Plots

plot	2-D line plot
plotyy	2-D line plots with y-axes on both left and right side
plot3	3-D line plot
loglog	Log-log scale plot
semilogx	Semilogarithmic plot
semilogy	Semilogarithmic plot
errorbar	Plot error bars along curve
fplot	Plot function between specified limits

ezplot	Easy-to-use function plotter
ezplot3	Easy-to-use 3-D parametric curve plotter
LineSpec (Line Specification)	Line specification string syntax
ColorSpec (Color Specification)	Color specification

2. Pie Charts, Bar Plots, and Histograms

bar	Bar graph
bar3	Plot 3-D bar graph
barh	Plot bar graph horizontally
bar3h	Plot horizontal 3-D bar graph
hist	Histogram plot
histc	Histogram bin count
rose	Angle histogram plot
pareto	Pareto chart
area	Filled area 2-D plot
pie	Pie chart
pie3	3-D pie chart

3. Discrete Data Plots

stem	Plot discrete sequence data
stairs	Stairstep graph
stem3	Plot 3-D discrete sequence data
scatter	Scatter plot
scatter3	3-D scatter plot
spy	Visualize sparsity pattern
plotmatrix	Scatter plot matrix

4. Polar Plots

polar	Polar coordinate plot
rose	Angle histogram plot
compass	Plot arrows emanating from origin
ezpolar	Easy-to-use polar coordinate plotter
LineSpec (Line Specification)	Line specification string syntax
ColorSpec (Color Specification)	Color specification

5. Contour Plots

contour	Contour plot of matrix
contourf	Filled 2-D contour plot
contourc	Low-level contour plot computation

contour3	3-D contour plot
contourslice	Draw contours in volume slice planes
ezcontour	Easy-to-use contour plotter
ezcontourf	Easy-to-use filled contour plotter

6. Vector Fields

feather	Plot velocity vectors
quiver	Quiver or velocity plot
compass	Plot arrows emanating from origin
quiver3	3-D quiver or velocity plot
streamslice	Plot streamlines in slice planes
streamline	Plot streamlines from 2-D or 3-D vector data

7. Surfaces, Volumes, and Polygons

1) Surface and Mesh Plots

surf	3-D shaded surface plot
surfc	Contour plot under a 3-D shaded surface plot
surface	Create surface object
surfl	Surface plot with colormap-based lighting
surfnorm	Compute and display 3-D surface normals
mesh	Mesh plot
meshc	Plot a contour graph under mesh graph
meshz	Plot a curtain around mesh plot
waterfall	Waterfall plot
ribbon	Ribbon plot
contour3	3-D contour plot
peaks	Example function of two variables
cylinder	Generate cylinder
ellipsoid	Generate ellipsoid
sphere	Generate sphere
pcolor	Pseudocolor (checkerboard) plot
surf2patch	Convert surface data to patch data
ezsurf	Easy-to-use 3-D colored surface plotter
ezsurfc	Easy-to-use combination surface/contour plotter
ezmesh	Easy-to-use 3-D mesh plotter
ezmeshc	Easy-to-use combination mesh/contour plotter

2) Volume Visualization

contourslice	Draw contours in volume slice planes

flow	Simple function of three variables
isocaps	Compute isosurface end-cap geometry
isocolors	Calculate isosurface and patch colors
isonormals	Compute normals of isosurface vertices
isosurface	Extract isosurface data from volume data
reducepatch	Reduce number of patch faces
reducevolume	Reduce number of elements in volume data set
shrinkfaces	Reduce size of patch faces
slice	Volumetric slice plots
mooth3	Smooth 3-D data
subvolume	Extract subset of volume data set
volumebounds	Coordinate and color limits for volume data
coneplot	Plot velocity vectors as cones in 3-D vector field
curl	Compute curl and angular velocity of vector field
divergence	Compute divergence of vector field
interpstreamspeed	Interpolate stream-line vertices from flow speed
stream2	Compute 2-D streamline data
stream3	Compute 3-D streamline data
streamline	Plot streamlines from 2-D or 3-D vector data
streamparticles	Plot stream particles
streamribbon	3-D stream ribbon plot from vector volume data
streamslice	Plot streamlines in slice planes
streamtube	Create 3-D stream tube plot

3) Polygons

fill	Filled 2-D polygons
fill3	Filled 3-D polygons
patch	Create one or more filled polygons
surf2patch	Convert surface data to patch data

8. Animation

movie	Play recorded movie frames
noanimate	Change Erase Mode of all objects to normal
drawnow	Update figure window and execute pending callbacks
refreshdata	Refresh data in graph when data source is specified
frame2im	Return image data associated with movie frame
getframe	Capture movie frame
im2frame	Convert image to movie frame
comet	2-D comet plot

comet3	3-D comet plot

I.3.2 Formatting and Annotation

1. Titles and Labels

title	Add title to current axes
xlabel	Label x-axis
ylabel	Label y-axis
zlabel	Label z-axis
clabel	Contour plot elevation labels
datetick	Date formatted tick labels
texlabel	Format text into TeX string
legend	Graph legend for lines and patches
colorbar	Colorbar showing color scale

2. Coordinate System

xlim	Set or query x-axis limits
ylim	Set or query y-axis limits
zlim	Set or query z-axis limits
box	Axes border
grid	Grid lines for 2-D and 3-D plots
daspect	Set or query axes data aspect ratio
pbaspect	Set or query plot box aspect ratio
axes	Create axes graphics object
axis	Axis scaling and appearance
subplot	Create axes in tiled positions
hold	Retain current graph when adding new graphs
gca	Current axes handle
cla	Clear current axes

3. Annotation

annotation	Create annotation objects
text	Create text object in current axes
legend	Graph legend for lines and patches
title	Add title to current axes
xlabel	Label x-axis
ylabel	Label y-axis
zlabel	Label z-axis

datacursormode	Enable, disable, and manage interactive data cursor mode
ginput	Graphical input from mouse or cursor
gtext	Mouse placement of text in 2-D view

4. Colormaps

colormap	Set and get current colormap
colormapeditor	Open colormap editor
colorbar	Colorbar showing color scale
brighten	Brighten or darken colormap
contrast	Grayscale colormap for contrast enhancement
shading	Set color shading properties
graymon	Set default figure properties for grayscale monitors
caxis	Color axis scaling
hsv2rgb	Convert HSV colormap to RGB colormap
rgb2hsv	Convert RGB colormap to HSV colormap
rgbplot	Plot colormap
spinmap	Spin colormap
colordef	Set default property values to display different color schemes
whitebg	Change axes background color

5. Data Exploration

hidden	Remove hidden lines from mesh plot
pan	Pan view of graph interactively
reset	Reset graphics object properties to their defaults
rotate	Rotate object about specified origin and direction
rotate3d	Rotate 3-D view using mouse
selectmoveresize	Select, move, resize, or copy axes and uicontrol graphics objects
zoom	Turn zooming on or off or magnify by factor Magnify by a factor
datacursormode	Enable, disable, and manage interactive data cursor mode
figurepalette	Show or hide Figure Palette
plotbrowser	Show or hide figure Plot Browser
plotedit	Interactively edit and annotate plots
plottools	Show or hide plot tools
propertyeditor	Show or hide Property Editor
showplottool	Show or hide figure plot tool

6. Data Brushing

brush	Interactively mark, delete, modify, and save observations in graphs
datacursormode	Enable, disable, and manage interactive data cursor mode
linkdata	Automatically update graphs when variables change
refreshdata	Refresh data in graph when data source is specified

I.3.3 Images

1. Image File Operations

image	Display image object
magesc	Scale data and display image object
imread	Read image from graphics file
imwrite	Write image to graphics file
imfinfo	Information about graphics file
imformats	Manage image file format registry
frame2im	Return image data associated with movie frame
im2frame	Convert image to movie frame
im2java	Convert image to Java image

2. Modifying Images

ind2rgb	Convert indexed image to RGB image
rgb2ind	Convert RGB image to indexed image
imapprox	Approximate indexed image by reducing number of colors
dither	Convert image, increasing apparent color resolution by dithering
cmpermute	Rearrange colors in colormap
cmunique	Eliminate duplicate colors in colormap; convert grayscale or truecolor image to indexed image

I.3.4 Printing and Exporting

print	Print figure or save to file
printopt	Configure printer defaults
printdlg	Print dialog box
printpreview	Preview figure to print
orient	Hardcopy paper orientation
savefig	Save figure to FIG-file
openfig	Open new copy or raise existing copy of saved figure

hgexport	Export figure
hgsave	Save Handle Graphics object hierarchy to file
hgload	Load Handle Graphics object hierarchy from file
saveas	Save figure or Simulink block diagram using specified format

I.3.5 Graphics Objects

1. Graphics Object Identification

gca	Current axes handle
gcf	Current figure handle
gcbf	Handle of figure containing object whose callback is executing
gcbo	Handle of object whose callback is executing
gco	Handle of current object
ancestor	Ancestor of graphics object
allchild	Find all children of specified objects
findall	Find all graphics objects
findfigs	Find visible offscreen figures
findobj	Locate graphics objects with specific properties
gobjects	Create array of graphics handles
ishghandle	True for Handle Graphics object handles
ishandle	Test for valid graphics or Java object handle
copyobj	Copy graphics objects and their descendants
delete	Remove files or objects
get	Query Handle Graphics object properties
set	Set Handle Graphics object properties
propedit	Open Property Editor

2. Core Objects

root object	Root
figure	Create figure graphics object
axes	Create axes graphics object
image	Display image object
light	Create light object
line	Create line object
patch	Create one or more filled polygons
rectangle	Create 2-D rectangle object

surface	Create surface object
text	Create text object in current axes

3. Annotation Objects

annotation	Create annotation objects

4. Plot Objects

set	Set Handle Graphics object properties
get	Query Handle Graphics object properties

5. Group Objects

hggroup	Create hggroup object
hgtransform	Create hgtransform graphics object
makehgtform	Create 4-by-4 transform matrix

6. Figure Windows

figure	Create figure graphics object
gcf	Current figure handle
close	Remove specified figure
clf	Clear current figure window
refresh	Redraw current figure
newplot	Determine where to draw graphics objects
shg	Show most recent graph window
closereq	Default figure close request function
dragrect	Drag rectangles with mouse
drawnow	Update figure window and execute pending callbacks
rbbox	Create rubberband box for area selection
opengl	Control OpenGL rendering

7. Axes Property Operations

axes	Create axes graphics object
hold	Retain current graph when adding new graphs
ishold	Current hold state
newplot	Determine where to draw graphics objects

8. Object Property Operations

linkaxes	Synchronize limits of specified 2-D axes
linkprop	Keep same value for corresponding properties of graphics objects
refreshdata	Refresh data in graph when data source is specified

waitfor	Block execution and wait for event or condition
get	Query Handle Graphics object properties
set	Set Handle Graphics object properties

I.4 Programming Scripts and Functions

I.4.1 Control Flow

if, elseif, else	Execute statements if condition is true
for	Execute statements specified number of times
parfor	Parallel for loop
switch, case, otherwise	Switch among several cases based on expression
try, catch	Execute statements and catch resulting errors
while	Repeatedly execute statements while condition is true
break	Terminate execution of for or while loop
continue	Pass control to next iteration of for or while loop
end	Terminate block of code, or indicate last array index
pause	Halt execution temporarily
return	Return to invoking function

I.4.2 Scripts

edit	Edit or create file
input	Request user input
publish	Generate view of MATLAB file in specified format
notebook	Open MATLAB Notebook in Microsoft Word software (on Microsoft Windows platforms)
grabcode	Extract MATLAB code from file published to HTML
snapnow	Force snapshot of image for inclusion in published document

I.4.3 Functions

1. Function Basics

function	Declare function name, inputs, and outputs

2. Input and Output Arguments

nargin	Number of function input arguments
nargout	Number of function output arguments

varargin	Variable-length input argument list
varargout	Variable-length output argument list
narginchk	Validate number of input arguments
nargoutchk	Validate number of output arguments
validateattributes	Check validity of array
validatestring	Check validity of text string
inputParser	Parse function inputs
inputname	Variable name of function input

3. Variables

persistent	Define persistent variable
isvarname	Determine whether input is valid variable name
matlab.lang.makeUniqueStrings	Construct unique strings from input strings
matlab.lang.makeValidName	Construct valid MATLAB identifiers from input strings
namelengthmax	Maximum identifier length
assignin	Assign value to variable in specified workspace
global	Declare global variables
isglobal	Determine whether input is global variable

4. Error Handling

try, catch	Execute statements and catch resulting errors
error	Display message and abort function
warning	Warning message
lastwarn	Last warning message
assert	Generate error when condition is violated
onCleanup	Cleanup tasks upon function completion

I.4.4 Debugging

dbclear	Clear breakpoints
dbcont	Resume execution
dbdown	Reverse workspace shift performed by dbup, while in debug mode
dbquit	Quit debug mode
dbstack	Function call stack
dbstatus	List all breakpoints
dbstep	Execute one or more lines from current breakpoint
dbstop	Set breakpoints for debugging
dbtype	List text file with line numbers

dbup	Shift current workspace to workspace of caller, while in debug mode
checkcode	Check MATLAB code files for possible problems
keyboard	Input from keyboard
mlintrpt	Run checkcode for file or folder, reporting results in browser

I.4.5 Coding and Productivity Tips

edit	Edit or create file

I.4.6 Programming Utilities

echo	Display statements during function execution
eval	Execute MATLAB expression in text string
evalc	Evaluate MATLAB expression with capture
evalin	Execute MATLAB expression in specified workspace
feval	Evaluate function
run	Run MATLAB script
builtin	Execute built-in function from overloaded method
matlab.codetools.requiredFilesAndProducts	List dependencies of MATLAB program files
mfilename	File name of currently running function
pcode	Create protected function file
timer	Create object to schedule execution of MATLAB commands

I.5 Data and File Management

I.5.1 Workspace Variables

clear	Remove items from workspace, freeing up system memory
clearvars	Clear variables from memory
disp	Display text or array
openvar	Open workspace variable in Variables editor or other graphical editing tool

who	List variables in workspace
whos	List variables in workspace, with sizes and types
load	Load variables from file into workspace
save	Save workspace variables to file
matfile	Access and change variables directly in MAT-files, without loading into memory

I.5.2 Data Import and Export

1. Import and Export Basics

importdata	Load data from file
uiimport	Import data interactively

2. Text Files

csvread	Read comma-separated value file
csvwrite	Write comma-separated value file
dlmread	Read ASCII-delimited file of numeric data into matrix
dlmwrite	Write matrix to ASCII-delimited file
textscan	Read formatted data from text file or string
readtable	Create table from file
writetable	Write table to file
type	Display contents of file

3. Spreadsheets

xlsfinfo	Determine if file contains Microsoft Excel spreadsheet
xlsread	Read Microsoft Excel spreadsheet file
xlswrite	Write Microsoft Excel spreadsheet file
readtable	Create table from file
writetable	Write table to file

4. Low-Level File I/O

fclose	Close one or all open files
feof	Test for end-of-file
ferror	Information about file I/O errors
fgetl	Read line from file, removing newline characters
fgets	Read line from file, keeping newline characters
fileread	Read contents of file into string
fopen	Open file, or obtain information about open files

fprintf	Write data to text file
fread	Read data from binary file
frewind	Move file position indicator to beginning of open file
fscanf	Read data from text file
fseek	Move to specified position in file
ftell	Position in open file
fwrite	Write data to binary file

5. Images

im2java	Convert image to Java image
imfinfo	Information about graphics file
imread	Read image from graphics file
imwrite	Write image to graphics file
Tiff	MATLAB Gateway to LibTIFF library routines

附录 II 图像处理函数

II.1 Import, Export, and Conversion

II.1.1 Basic Import and Export

imread	Read image from graphics file
imwrite	Write image to graphics file
imfinfo	Information about graphics file

II.1.2 Scientific File Formats

dicomanon	Anonymize DICOM file
dicomdict	Get or set active DICOM data dictionary
dicominfo	Read metadata from DICOM message
dicomlookup	Find attribute in DICOM data dictionary
dicomread	Read DICOM image
dicomuid	Generate DICOM unique identifier
dicomwrite	Write images as DICOM files
nitfinfo	Read metadata from National Imagery Transmission Format (NITF) file
nitfread	Read image from NITF file
analyze75info	Read metadata from header file of Analyze 7.5 data set
analyze75read	Read image data from image file of Analyze 7.5 data set
interfileinfo	Read metadata from Interfile file
interfileread	Read images in Interfile format

II.1.3 High Dynamic Range Images

hdrread	Read high dynamic range (HDR) image
hdrwrite	Write Radiance high dynamic range (HDR) image file
makehdr	Create high dynamic range image
tonemap	Render high dynamic range image for viewing

II.1.4 Large Image Files

ImageAdapter	Interface for image I/O
isrset	Check if file is R-Set

openrset	Open R-Set file
rsetwrite	Create reduced resolution data set from image file

II.1.5 Image Type Conversion

gray2ind	Convert grayscale or binary image to indexed image
ind2gray	Convert indexed image to grayscale image
mat2gray	Convert matrix to grayscale image
rgb2gray	Convert RGB image or colormap to grayscale
ind2rgb	Convert indexed image to RGB image
label2rgb	Convert label matrix into RGB image
demosaic	Convert Bayer pattern encoded image to truecolor image
imquantize	Quantize image using specified quantization levels and output values
multithresh	Multilevel image thresholds using Otsu's method
im2bw	Convert image to binary image, based on threshold
graythresh	Global image threshold using Otsu's method
grayslice	Convert grayscale image to indexed image using multilevel thresholding
im2double	Convert image to double precision
im2int16	Convert image to 16-bit signed integers
im2java2d	Convert image to Java buffered image
im2single	Convert image to single precision
im2uint16	Convert image to 16-bit unsigned integers
im2uint8	Convert image to 8-bit unsigned integers

II.1.6 Synthetic Images

checkerboard	Create checkerboard image
phantom	Create head phantom image
imnoise	Add noise to image

II.2 Display and Exploration

II.2.1 Basic Display

imshow	Display image
montage	Display multiple image frames as rectangular montage
subimage	Display multiple images in single figure
immovie	Make movie from multiframe image

implay	Play movies, videos, or image sequences
warp	Display image as texture-mapped surface
iptgetpref	Get values of Image Processing Toolbox preferences
iptprefs	Display Image Processing Preferences dialog box
iptsetpref	Set Image Processing Toolbox preferences or display valid values

II.2.2　Interactive Exploration with the Image Viewer App

imtool	Image Viewer app
imageinfo	Image Information tool
imcontrast	Adjust Contrast tool
imdisplayrange	Display Range tool
imdistline	Distance tool
impixelinfo	Pixel Information tool
impixelinfoval	Pixel Information tool without text label
impixelregion	Pixel Region tool
immagbox	Magnification box for scroll panel
imoverview	Overview tool for image displayed in scroll panel
iptgetpref	Get values of Image Processing Toolbox preferences
iptprefs	Display Image Processing Preferences dialog box
iptsetpref	Set Image Processing Toolbox preferences or display valid values

II.2.3　Build Interactive Tools

imageinfo	Image Information tool
imcolormaptool	Choose Colormap tool
imcontrast	Adjust Contrast tool
imcrop	Crop image
imdisplayrange	Display Range tool
imdistline	Distance tool
impixelinfo	Pixel Information tool
impixelinfoval	Pixel Information tool without text label
impixelregion	Pixel Region tool
impixelregionpanel	Pixel Region tool panel
immagbox	Magnification box for scroll panel
imoverview	Overview tool for image displayed in scroll panel
imoverviewpanel	Overview tool panel for image displayed in scroll panel
imsave	Save Image Tool

imscrollpanel	Scroll panel for interactive image navigation
imellipse	Create draggable ellipse
imfreehand	Create draggable freehand region
imline	Create draggable, resizable line
impoint	Create draggable point
impoly	Create draggable, resizable polygon
imrect	Create draggable rectangle
imroi	Region-of-interest (ROI) base class
getline	Select polyline with mouse
getpts	Specify points with mouse
getrect	Specify rectangle with mouse
getimage	Image data from axes
getimagemodel	Image model object from image object
axes2pix	Convert axes coordinates to pixel coordinates
imattributes	Information about image attributes
imgca	Get handle to current axes containing image
imgcf	Get handle to current figure containing image
imgetfile	Open Image dialog box
imhandles	Get all image handles
iptaddcallback	Add function handle to callback list
iptcheckhandle	Check validity of handle
iptgetapi	Get Application Programmer Interface (API) for handle
iptGetPointerBehavior	Retrieve pointer behavior from HG object
ipticondir	Directories containing IPT and MATLAB icons
iptPointerManager	Create pointer manager in figure
iptremovecallback	Delete function handle from callback list
iptSetPointerBehavior	Store pointer behavior structure in Handle Graphics object
iptwindowalign	Align figure windows
makeConstrainToRectFcn	Create rectangularly bounded drag constraint function
truesize	Adjust display size of image

II.3 Geometric Transformation, Spatial Referencing, and Image Registration

II.3.1 Geometric Transformations

imcrop	Crop image
imresize	Resize image

imrotate	Rotate image
imtranslate	Translate image
impyramid	Image pyramid reduction and expansion
imwarp	Apply geometric transformation to image
fitgeotrans	Fit geometric transformation to control point pairs
imtransform	Apply 2-D spatial transformation to image
findbounds	Find output bounds for spatial transformation
fliptform	Flip input and output roles of TFORM structure
makeresampler	Create resampling structure
maketform	Create spatial transformation structure（TFORM）
tformarray	Apply spatial transformation to N-D array
tformfwd	Apply forward spatial transformation
tforminv	Apply inverse spatial transformation
checkerboard	Create checkerboard image
affine2d	2-D Affine Geometric Transformation
affine3d	3-D Affine Geometric Transformation
projective2d	2-D Projective Geometric Transformation
images.geotrans.PiecewiseLinearTransformation2D	2D piecewise linear geometric transformation
images.geotrans.PolynomialTransformation2D	2D Polynomial Geometric Transformation
images.geotrans.LocalWeightedMeanTransformation2D	2D Local Weighted Mean Geometric Transformation

II.3.2 Spatial Referencing

imwarp	Apply geometric transformation to image
imregister	Intensity-based image registration
imregtform	Estimate geometric transformation that aligns two 2-D or 3-D images
imshow	Display image
imshowpair	Compare differences between images
imfuse	Composite of two images
imref2d	Reference 2-D image to world coordinates
imref3d	Reference 3-D image to world coordinates

II.3.3 Automatic Registration

imregister	Intensity-based image registration
imregconfig	Configurations for intensity-based registration

imregtform	Estimate geometric transformation that aligns two 2-D or 3-D images
imregcorr	Estimates geometric transformation that aligns two 2-D images using phase correlation
imfuse	Composite of two images
imshowpair	Compare differences between images
registration.metric.MattesMutualInformation	Mattes mutual information metric configuration object
registration.metric.MeanSquares	Mean square error metric configuration object
registration.optimizer.RegularStepGradientDescent	Regular step gradient descent optimizer configuration object
registration.optimizer.OnePlusOneEvolutionary	One-plus-one evolutionary optimizer configuration object

II.3.4　Control Point Registration

cpselect	Control Point Selection Tool
fitgeotrans	Fit geometric transformation to control point pairs
cpcorr	Tune control-point locations using cross correlation
cpstruct2pairs	Convert CPSTRUCT to valid pairs of control points
normxcorr2	Normalized 2-D cross-correlation
cp2tform	Infer spatial transformation from control point pairs

II.4　Image Enhancement

II.4.1　Contrast Adjustment

imadjust	Adjust image intensity values or colormap
imcontrast	Adjust Contrast tool
imsharpen	Sharpen image using unsharp masking
histeq	Enhance contrast using histogram equalization
adapthisteq	Contrast-limited adaptive histogram equalization (CLAHE)
imhistmatch	Adjust histogram of image to match N-bin histogram of reference image
decorrstretch	Apply decorrelation stretch to multichannel images
tretchlim	Find limits to contrast stretch image
intlut	Convert integer values using lookup table
imnoise	Add noise to image

II.4.2　Image Filtering

imfilter	N-D filtering of multidimensional images
normxcorr2	Normalized 2-D cross-correlation
fspecial	Create predefined 2-D filter
wiener2	2-D adaptive noise-removal filtering
padarray	Pad array
freqz2	2-D frequency response
fsamp2	2-D FIR filter using frequency sampling
ftrans2	2-D FIR filter using frequency transformation
fwind1	2-D FIR filter using 1-D window method
fwind2	2-D FIR filter using 2-D window method
convmtx2	2-D convolution matrix
imguidedfilter	Guided filtering of images
nlfilter	General sliding-neighborhood operations
medfilt2	2-D median filtering
ordfilt2	2-D order-statistic filtering
stdfilt	Local standard deviation of image
rangefilt	Local range of image
entropyfilt	Local entropy of grayscale image

II.4.3　Morphological Operations

bwhitmiss	Binary hit-miss operation
bwmorph	Morphological operations on binary images
bwulterode	Ultimate erosion
bwareaopen	Remove small objects from binary image
imbothat	Bottom-hat filtering
imclearborder	Suppress light structures connected to image border
imclose	Morphologically close image
imdilate	Dilate image
imerode	Erode image
imextendedmax	Extended-maxima transform
imextendedmin	Extended-minima transform
imfill	Fill image regions and holes
imhmax	H-maxima transform
imhmin	H-minima transform
imimposemin	Impose minima
imopen	Morphologically open image

imreconstruct	Morphological reconstruction
imregionalmax	Regional maxima
imregionalmin	Regional minima
imtophat	Top-hat filtering
watershed	Watershed transform
conndef	Create connectivity array
iptcheckconn	Check validity of connectivity argument
applylut	Neighborhood operations on binary images using lookup tables
bwlookup	Nonlinear filtering using lookup tables
makelut	Create lookup table for use with bwlookup
strel	Create morphological structuring element (STREL)
getheight	Height of structuring element
getneighbors	Structuring element neighbor locations and heights
getnhood	Structuring element neighborhood
getsequence	Sequence of decomposed structuring elements
isflat	True for flat structuring element
reflect	Reflect structuring element
translate	Translate structuring element (STREL)

II.4.4 Deblurring

deconvblind	Deblur image using blind deconvolution
deconvlucy	Deblur image using Lucy-Richardson method
deconvreg	Deblur image using regularized filter
deconvwnr	Deblur image using Wiener filter
edgetaper	Taper discontinuities along image edges
otf2psf	Convert optical transfer function to point-spread function
psf2otf	Convert point-spread function to optical transfer function
padarray	Pad array

II.4.5 ROI-Based Processing

roipoly	Specify polygonal region of interest (ROI)
poly2mask	Convert region of interest (ROI) polygon to region mask
roicolor	Select region of interest (ROI) based on color
roifill	Fill in specified region of interest (ROI) polygon in grayscale image
roifilt2	Filter region of interest (ROI) in image
imellipse	Create draggable ellipse

imfreehand	Create draggable freehand region
impoly	Create draggable, resizable polygon
imrect	Create draggable rectangle
imroi	Region-of-interest (ROI) base class

II.4.6 Neighborhood and Block Processing

ImageAdapter	Interface for image I/O
blockproc	Distinct block processing for image
bestblk	Determine optimal block size for block processing
nlfilter	General sliding-neighborhood operations
col2im	Rearrange matrix columns into blocks
colfilt	Columnwise neighborhood operations
im2col	Rearrange image blocks into columns

II.4.7 Image Arithmetic

imabsdiff	Absolute difference of two images
imadd	Add two images or add constant to image
imapplymatrix	Linear combination of color channels
imcomplement	Complement image
imdivide	Divide one image into another or divide image by constant
imlincomb	Linear combination of images
immultiply	Multiply two images or multiply image by constant
imsubtract	Subtract one image from another or subtract constant from image

II.5 Image Analysis

II.5.1 Object Analysis

bwboundaries	Trace region boundaries in binary image
bwtraceboundary	Trace object in binary image
corner	Find corner points in image
cornermetric	Create corner metric matrix from image
edge	Find edges in intensity image
hough	Hough transform
houghlines	Extract line segments based on Hough transform
houghpeaks	Identify peaks in Hough transform
imfindcircles	Find circles using circular Hough transform

imgradient	Gradient magnitude and direction of an image
imgradientxy	Directional gradients of an image
viscircles	Create circle
qtdecomp	Quadtree decomposition
qtgetblk	Block values in quadtree decomposition
qtsetblk	Set block values in quadtree decomposition

II.5.2 Region and Image Properties

regionprops	Measure properties of image regions
bwarea	Area of objects in binary image
bwconncomp	Find connected components in binary image
bwconvhull	Generate convex hull image from binary image
bwdist	Distance transform of binary image
bwdistgeodesic	Geodesic distance transform of binary image
bweuler	Euler number of binary image
bwperim	Find perimeter of objects in binary image
bwselect	Select objects in binary image
graydist	Gray-weighted distance transform of grayscale image
imcontour	Create contour plot of image data
imhist	Histogram of image data
impixel	Pixel color values
improfile	Pixel-value cross-sections along line segments
corr2	2-D correlation coefficient
mean2	Average or mean of matrix elements
std2	Standard deviation of matrix elements
bwlabel	Label connected components in 2-D binary image
bwlabeln	Label connected components in binary image
labelmatrix	Create label matrix from bwconncomp structure
bwpack	Pack binary image
bwunpack	Unpack binary image

II.5.3 Texture Analysis

entropy	Entropy of grayscale image
entropyfilt	Local entropy of grayscale image
rangefilt	Local range of image
stdfilt	Local standard deviation of image
graycomatrix	Create gray-level co-occurrence matrix from image
graycoprops	Properties of gray-level co-occurrence matrix

II.5.4　Image Quality

ssim	Structural Similarity Index (SSIM) for measuring image quality
psnr	Peak Signal-to-Noise Ratio (PSNR)

II.5.5　Image Segmentation

activecontour	Segment image into foreground and background using active contour
graythresh	Global image threshold using Otsu's method
multithresh	Multilevel image thresholds using Otsu's method
colorThresholder	Threshold color image

II.5.6　Image Transforms

bwdist	Distance transform of binary image
bwdistgeodesic	Geodesic distance transform of binary image
graydist	Gray-weighted distance transform of grayscale image
hough	Hough transform
dct2	2-D discrete cosine transform
dctmtx	Discrete cosine transform matrix
fan2para	Convert fan-beam projections to parallel-beam
fanbeam	Fan-beam transform
idct2	2-D inverse discrete cosine transform
ifanbeam	Inverse fan-beam transform
iradon	Inverse Radon transform
para2fan	Convert parallel-beam projections to fan-beam
radon	Radon transform
fft2	2-D fast Fourier transform
fftshift	Shift zero-frequency component to center of spectrum
ifft2	2-D inverse fast Fourier transform
ifftshift	Inverse FFT shift

II.6　Color

makecform	Create color transformation structure
applycform	Apply device-independent color space transformation
iccfind	Search for ICC profiles
iccread	Read ICC profile

iccroot	Find system default ICC profile repository
iccwrite	Write ICC color profile to disk file
isicc	True for valid ICC color profile
imapprox	Approximate indexed image by reducing number of colors
lab2double	Convert L*a*b* data to double
lab2uint16	Convert L*a*b* data to uint16
lab2uint8	Convert L*a*b* data to uint8
ntsc2rgb	Convert NTSC values to RGB color space
rgb2ntsc	Convert RGB color values to NTSC color space
rgb2ycbcr	Convert RGB color values to YCbCr color space
xyz2double	Convert XYZ color values to double
xyz2uint16	Convert XYZ color values to uint16
ycbcr2rgb	Convert YCbCr color values to RGB color space
whitepoint	XYZ color values of standard illuminants

II.7　Code Generation

imfilter	N-D filtering of multidimensional images
imhist	Histogram of image data
fspecial	Create predefined 2-D filter
edge	Find edges in intensity image
mean2	Average or mean of matrix elements
imwarp	Apply geometric transformation to image
label2rgb	Convert label matrix into RGB image
bwlookup	Nonlinear filtering using lookup tables
bwselect	Select objects in binary image
im2double	Convert image to double precision
im2int16	Convert image to 16-bit signed integers
im2single	Convert image to single precision
im2uint16	Convert image to 16-bit unsigned integers
im2uint8	Convert image to 8-bit unsigned integers
bwmorph	Morphological operations on binary images
imbothat	Bottom-hat filtering
imclose	Morphologically close image
imdilate	Dilate image
imerode	Erode image
imextendedmax	Extended-maxima transform
imextendedmin	Extended-minima transform

imfill	Fill image regions and holes
imhmax	H-maxima transform
imhmin	H-minima transform
imopen	Morphologically open image
imreconstruct	Morphological reconstruction
imregionalmax	Regional maxima
imregionalmin	Regional minima
imtophat	Top-hat filtering
bwpack	Pack binary image
bwunpack	Unpack binary image
conndef	Create connectivity array
getrangefromclass	Default display range of image based on its class
imcomplement	Complement image
iptcheckconn	Check validity of connectivity argument
padarray	Pad array
strel	Create morphological structuring element (STREL)
imref2d	Reference 2-D image to world coordinates
imref3d	Reference 3-D image to world coordinates
affine2d	2-D Affine Geometric Transformation
projective2d	2-D Projective Geometric Transformation

附录 III 小波分析函数

III.1 Wavelets and Filter Banks

III.1.1 Real and Complex-Valued Wavelets

bswfun	Biorthogonal scaling and wavelet functions
centfrq	Wavelet center frequency
cgauwavf	Complex Gaussian wavelet
cmorwavf	Complex Morlet wavelet
fbspwavf	Complex frequency B-spline wavelet
gauswavf	Gaussian wavelet
intwave	Integrate wavelet function psi (ψ)
mexihat	Mexican hat wavelet
meyer	Meyer wavelet
meyeraux	Meyer wavelet auxiliary function
morlet	Morlet wavelet
scal2frq	Scale to frequency
shanwavf	Complex Shannon wavelet
wavefun	Wavelet and scaling functions
wavefun2	Wavelet and scaling functions 2-D
wavsupport	Wavelet support
wavemenu	Wavelet Toolbox GUI tools
wavemngr	Wavelet manager
waveletfamilies	Wavelet families and family members
waveinfo	Wavelets information
wavedemo	Wavelet Toolbox software examples
wtbxmngr	Wavelet Toolbox manager

III.1.2 Orthogonal and Biorthogonal Filter Banks

biorwavf	Biorthogonal spline wavelet filter
biorfilt	Biorthogonal wavelet filter set
coifwavf	Coiflet wavelet filter
dddtree	Dual-tree and double-density 1-D wavelet transform

dbaux	Daubechies wavelet filter computation
dbwavf	Daubechies wavelet filter
orthfilt	Orthogonal wavelet filter set
rbiowavf	Reverse biorthogonal spline wavelet filters
qmf	Scaling and Wavelet Filter
symaux	Symlet wavelet filter computation
symwavf	Symlet wavelet filter
wavefun	Wavelet and scaling functions
wavefun2	Wavelet and scaling functions 2-D
wfilters	Wavelet filters
wrev	Flip vector
wavemenu	Wavelet Toolbox GUI tools
wavemngr	Wavelet manager
waveletfamilies	Wavelet families and family members
waveinfo	Wavelets information
wavenames	Wavelet names for LWT

III.1.3 Lifting

addlift	Add lifting steps to lifting scheme
displs	Display lifting scheme
filt2ls	Transform quadruplet of filters to lifting scheme
laurmat	Laurent matrices constructor
laurpoly	Laurent polynomials constructor
liftfilt	Apply elementary lifting steps on quadruplet of filters
liftwave	Lifting schemes
lsinfo	Lifting schemes information
ls2filt	Transform lifting scheme to quadruplet of filters
wave2lp	Laurent polynomials associated with wavelet
wavemngr	Wavelet manager
waveletfamilies	Wavelet families and family members
waveinfo	Wavelets information
wavenames	Wavelet names for LWT

III.1.4 Wavelet Design

pat2cwav	Build wavelet from pattern
wavemenu	Wavelet Toolbox GUI tools
wavemngr	Wavelet manager
addlift	Add lifting steps to lifting scheme

displs	Display lifting scheme
filt2ls	Transform quadruplet of filters to lifting scheme
laurmat	Laurent matrices constructor
laurpoly	Laurent polynomials constructor
liftfilt	Apply elementary lifting steps on quadruplet of filters
liftwave	Lifting schemes
lsinfo	Lifting schemes information
ls2filt	Transform lifting scheme to quadruplet of filters
wave2lp	Laurent polynomials associated with wavelet

III.2　Continuous Wavelet Analysis

conofinf	Cone of influence
cwt	Continuous 1-D wavelet transform
cwtext	Real or complex continuous 1-D wavelet coefficients using extension parameters
cwtft	Continuous wavelet transform using FFT algorithm
cwtftinfo	Valid analyzing wavelets for FFT-based CWT
cwtftinfo2	Supported 2-D CWT wavelets and Fourier transforms
cwtft2	2-D continuous wavelet transform
icwtft	Inverse CWT
icwtlin	Inverse continuous wavelet transform（CWT）for linearly spaced scales
localmax	Identify and chain local maxima
pat2cwav	Build wavelet from pattern
wcoher	Wavelet coherence
wscalogram	Scalogram for continuous wavelet transform
wavemenu	Wavelet Toolbox GUI tools
wavemngr	Wavelet manager

III.3　Discrete Wavelet Analysis

III.3.1　Signal Analysis

dyaddown	Dyadic downsampling
dyadup	Dyadic upsampling
upcoef	Direct reconstruction from 1-D wavelet coefficients
appcoef	1-D approximation coefficients

detcoef	1-D detail coefficients
wrcoef	Reconstruct single branch from 1-D wavelet coefficients
dwt	Single-level discrete 1-D wavelet transform
dwtmode	Discrete wavelet transform extension mode
idwt	Single-level inverse discrete 1-D wavelet transform
waverec	Multilevel 1-D wavelet reconstruction
wavedec	Multilevel 1-D wavelet decomposition
upwlev	Single-level reconstruction of 1-D wavelet decomposition
lwt	1-D lifting wavelet transform
lwtcoef	Extract or reconstruct 1-D LWT wavelet coefficients
ilwt	Inverse 1-D lifting wavelet transform
swt	Discrete stationary wavelet transform 1-D
iswt	Inverse discrete stationary wavelet transform 1-D
ndwt	Nondecimated 1-D wavelet transform
indwt	Inverse nondecimated 1-D wavelet transform
dddtree	Dual-tree and double-density 1-D wavelet transform
dddtreecfs	Extract dual-tree/double-density wavelet coefficients or projections
dtfilters	Analysis and synthesis filters for oversampled wavelet filter banks
idddtree	Inverse dual-tree and double-density 1-D wavelet transform
plotdt	Plot dual-tree or double-density wavelet transform
wenergy	Energy for 1-D wavelet or wavelet packet decomposition
wvarchg	Find variance change points
wmaxlev	Maximum wavelet decomposition level
wfbm	Fractional Brownian motion synthesis
wfbmesti	Parameter estimation of fractional Brownian motion
measerr	Approximation quality metrics
wrev	Flip vector
wextend	Extend vector or matrix
wkeep	Keep part of vector or matrix
wavemenu	Wavelet Toolbox GUI tools
wavemngr	Wavelet manager

III.3.2 Image Analysis

dyaddown	Dyadic downsampling
dyadup	Dyadic upsampling
upcoef2	Direct reconstruction from 2-D wavelet coefficients

appcoef2	2-D approximation coefficients
detcoef2	2-D detail coefficients
dwt2	Single-level discrete 2-D wavelet transform
dwtmode	Discrete wavelet transform extension mode
idwt2	Single-level inverse discrete 2-D wavelet transform
wavedec2	Multilevel 2-D wavelet decomposition
waverec2	Multilevel 2-D wavelet reconstruction
wrcoef2	Reconstruct single branch from 2-D wavelet coefficients
upwlev2	Single-level reconstruction of 2-D wavelet decomposition
wenergy2	Energy for 2-D wavelet decomposition
ilwt2	Inverse 2-D lifting wavelet transform
iswt2	Inverse discrete stationary wavelet transform 2-D
lwt2	2-D lifting wavelet transform
lwtcoef2	Extract or reconstruct 2-D LWT wavelet coefficients
swt2	Discrete stationary wavelet transform 2-D
iswt2	Inverse discrete stationary wavelet transform 2-D
ndwt2	Nondecimated 2-D wavelet transform
indwt2	Inverse nondecimated 2-D wavelet transform
dddtreecfs	Extract dual-tree/double-density wavelet coefficients or projections
dddtree2	Dual-tree and double-density 2-D wavelet transform
dtfilters	Analysis and synthesis filters for oversampled wavelet filter banks
idddtree2	Inverse dual-tree and double-density 2-D wavelet transform
plotdt	Plot dual-tree or double-density wavelet transform
wcodemat	Extended pseudocolor matrix scaling
wfusimg	Fusion of two images
wfusmat	Fusion of two matrices or arrayz
measerr	Approximation quality metrics
wextend	Extend vector or matrix
wkeep	Keep part of vector or matrix
wavemenu	Wavelet Toolbox GUI tools
wavemngr	Wavelet manager

III.3.3　3-D Analysis

dwt3	Single-level discrete 3-D wavelet transform
dwtmode	Discrete wavelet transform extension mode
idwt3	Single-level inverse discrete 3-D wavelet transform

wavedec3	Multilevel 3-D wavelet decomposition
waverec3	Multilevel 3-D wavelet reconstruction
wavemenu	Wavelet Toolbox GUI tools
wavemngr	Wavelet manager

III.3.4 Multisignal Analysis

chgwdeccfs	Change multisignal 1-D decomposition coefficients
dwtmode	Discrete wavelet transform extension mode
mdwtcluster	Multisignals 1-D clustering
mdwtdec	Multisignal 1-D wavelet decomposition
mdwtrec	Multisignal 1-D wavelet reconstruction
mswcmp	Multisignal 1-D compression using wavelets
mswcmpscr	Multisignal 1-D wavelet compression scores
mswcmptp	Multisignal 1-D compression thresholds and performances
mswden	Multisignal 1-D denoising using wavelets
mswthresh	Perform multisignal 1-D thresholding
wavemenu	Wavelet Toolbox GUI tools
wdecenergy	Multisignal 1-D decomposition energy distribution
wmspca	Multiscale Principal Component Analysis
wextend	Extend vector or matrix
wkeep	Keep part of vector or matrix
wavemenu	Wavelet Toolbox GUI tools

III.4 Wavelet Packet Analysis

dwtmode	Discrete wavelet transform extension mode
wpdec	Wavelet packet decomposition 1-D
wpdec2	Wavelet packet decomposition 2-D
wprec	Wavelet packet reconstruction 1-D
wprec2	Wavelet packet reconstruction 2-D
wpcoef	Wavelet packet coefficients
wprcoef	Reconstruct wavelet packet coefficients
bestlevt	Best level tree wavelet packet analysis
besttree	Best tree wavelet packet analysis
entrupd	Entropy update (wavelet packet)
wentropy	Entropy (wavelet packet)
plot	Plot tree GUI

wpviewcf	Plot wavelet packets colored coefficients
wavemenu	Wavelet Toolbox GUI tools
wpfun	Wavelet packet functions
wpspectrum	Wavelet packet spectrum
cfs2wpt	Wavelet packet tree construction from coefficients
depo2ind	Node depth-position to node index
wp2wtree	Extract wavelet tree from wavelet packet tree
wpcutree	Cut wavelet packet tree
wpsplt	Split (decompose) wavelet packet
wpjoin	Recompose wavelet packet
ind2depo	Node index to node depth-position
otnodes	Order terminal nodes of binary wavelet packet tree
write	Write values in WPTREE fields
read	Read values of WPTREE
readtree	Read wavelet packet decomposition tree from figure
set	WPTREE field contents
tnodes	Determine terminal nodes
wptree	WPTREE constructor
disp	WPTREE information
drawtree	Draw wavelet packet decomposition tree (GUI)
dtree	DTREE constructor
allnodes	Tree nodes
get	WPTREE contents
isnode	Existing node test
istnode	Terminal nodes indices test
leaves	Determine terminal nodes
nodeasc	Node ascendants
nodedesc	Node descendants
nodejoin	Recompose node
nodepar	Node parent
nodesplt	Split (decompose) node
noleaves	Determine nonterminal nodes
ntnode	Number of terminal nodes
ntree	NTREE constructor
treedpth	Tree depth
treeord	Tree order
wtbo	WTBO constructor
wtreemgr	NTREE manager

III.5 Denoising

cmddenoise	Interval-dependent denoising
ddencmp	Default values for denoising or compression
thselect	Threshold selection for de-noising
wbmpen	Penalized threshold for wavelet 1-D or 2-D de-noising
wdcbm	Thresholds for wavelet 1-D using Birgé-Massart strategy
wdcbm2	Thresholds for wavelet 2-D using Birgé-Massart strategy
wden	Automatic 1-D de-noising
wdencmp	De-noising or compression
wmulden	Wavelet multivariate de-noising
wnoise	Noisy wavelet test data
wnoisest	Estimate noise of 1-D wavelet coefficients
wpbmpen	Penalized threshold for wavelet packet de-noising
wpdencmp	De-noising or compression using wavelet packets
wpthcoef	Wavelet packet coefficients thresholding
wthcoef	1-D wavelet coefficient thresholding
wthcoef2	Wavelet coefficient thresholding 2-D
wthresh	Soft or hard thresholding
wthrmngr	Threshold settings manager
wvarchg	Find variance change points
measerr	Approximation quality metrics
wavemenu	Wavelet Toolbox GUI tools

参 考 文 献

曹茂永. 2007. 数字图像处理[M]. 北京: 北京大学出版社
葛哲学, 沙威. 2008. 小波分析理论与 MATLAB R2007 实现[M]. 北京: 电子工业出版社
郭文强, 侯勇严. 2009. 数字图像处理[M]. 西安: 西安电子科技大学出版社
韩晓军. 2009. 数字图像处理技术及应用[M]. 北京: 电子工业出版社
侯宏花. 2011. 数字图像处理与分析[M]. 北京: 北京理工大学出版社
胡学龙. 2011. 数字图像处理[M]. 北京: 电子工业出版社
李俊山, 李旭辉. 2007. 数字图像处理[M]. 北京: 清华大学出版社
阮秋琦. 2013. 数字图像处理学[M]. 北京: 电子工业出版社
王爱玲, 叶明生, 邓秋香. 2008. MATLAB R2007 图像处理技术与应用[M]. 北京: 电子工业出版社
王开福. 2013. 现代光测及其图像处理[M]. 北京: 科学出版社
王志明, 殷绪成, 曾慧. 2012. 数字图像处理与分析[M]. 北京: 清华大学出版社
谢凤英, 赵丹培, 李露, 等. 2014. 数字图像处理及应用[M]. 北京: 电子工业出版社
闫敬文. 2011. 数字图像处理(MATLAB 版)[M]. 北京: 国防工业出版社
姚峰林. 2014. 数字图像处理及在工程中的应用[M]. 北京: 北京理工大学出版社
张岩. 2014. MATLAB 图像处理超级学习手册[M]. 北京: 人民邮电出版社
章毓晋. 2012. 图像工程[M]. 北京: 清华大学出版社